Palgrave Studies in the History of Economic Thought

Palgrave Studies in the History of Economic Thought publishes contributions by leading scholars, illuminating key events, theories and individuals that have had a lasting impact on the development of modern-day economics. The topics covered include the development of economies, institutions and theories.

More information about this series at
http://www.palgrave.com/gp/series/14585

Renata Allio

# War in Economic Theories over Time

## Assessing the True Economic, Social and Political Costs

palgrave
macmillan

Renata Allio
University of Turin
Turin, Italy

ISSN 2662-6578 ISSN 2662-6586 (electronic)
Palgrave Studies in the History of Economic Thought
ISBN 978-3-030-39616-9 ISBN 978-3-030-39617-6 (eBook)
https://doi.org/10.1007/978-3-030-39617-6

Translation from the Italian language edition *Gli economisti e la Guerra* by Renata Allio, 

Cover illustration: Filo/Getty Images

This Palgrave Macmillan imprint is published by the registered company Springer Nature Switzerland AG
The registered company address is: Gewerbestrasse 11, 6330 Cham, Switzerland

# Acknowledgements

I would like to thank above all Palgrave Macmillan who generously extended the time allowed for the publication of this work to offer me the opportunity, after a period of illness, to make the necessary revision and update the Italian edition, published late in 2014. I would also like to thank the editorial staff who followed the publication through its various phases with courtesy and impeccably professional work. Thanks should also go to David Brown for his indispensable help in the translation from Italian and to Giulia Rubino who patiently and carefully revised the text and the drafts and compiled the index.

# Contents


| | | |
|---|---|---|
| 1 | Introduction | 1 |
| 2 | War and Economic Activity | 13 |
| 3 | From Historicizing to the Obsolescence of War | 57 |
| 4 | Imperialism | 77 |
| 5 | The Cost of War | 101 |
| 6 | Economic Rationality or Irrelevance of War? | 133 |
| 7 | The Benefits of War and the Armaments Industry | 171 |
| 8 | Economic Globalization, Realpolitik, New Wars | 217 |
| 9 | Conclusions | 245 |
| Index | | 259 |

# 1
# Introduction

What are the relationships between war and economic activity? When do either public or private economic requirements cause war? Are war-like conflicts rational from an economic point of view? What is the best strategy to wage them? What are the economic consequences of war? How much do wars really cost? Who really gains from them? Can war or forceful rearmament contribute to resolving economic crises?

Between the two far apart historical poles of mercantilism and mathematical economists of Game theory, separated by 450 years, different schools and individual students have taken into consideration various aspects of waging war and the reasons for it, following the stimuli offered by the problems cropping up over time that came to the forefront with greater urgency. The list includes: aggressive versus pacific foreign trade, the need for lasting peace in order to establish and sustain industrial development, the need for war to conclude processes of political and economic unification, rearmament and colonialization in order to overcome economic crises and stagnation, imperialism, the costs of world wars and the way to cover them, wartime financing, the strategies to be applied when there is the possibility of devastating nuclear conflicts and the role of deterrence in favouring peace.



R. Allio, *War in Economic Theories over Time*,
Palgrave Studies in the History of Economic Thought,
https://doi.org/10.1007/978-3-030-39617-6_1

Let us look briefly at the historical panorama: between the late 1500s and the 1600s, the mercantilists sought to define the links between economic logic and war. The power and enrichment of the state were then considered as the central problem of the economy, so, seeing that it was difficult to produce new wealth, it appeared easier to attempt to seize that of other countries. Consequently, wars were the indispensable and normal instrument to reach the goals of the economy itself, and turned out to be a zero-sum game as what one contender gained, the other lost.

Later on, in the period between the late 1700s and the early 1800s, the rise of the process of industrialization, which made it possible to raise production significantly, the English and French free-traders made individual well-being the object of economic strength. This well-being was realized with productive competition in a state of free trade and international peace. Since the wealth of the state was nothing else than the sum of the wealth of the citizens, individual well-being coincided with that of the state. This explains why the free traders paid little attention to the causes and the effects of warlike phenomena, which, as they saw it, were destined to disappear with progress in economic development. In the immediate future, the only war that could be envisaged was one of defence, and the state had to deal with this. Peace had to be sought after actively: indeed certain French and English free traders in the late 1800s and early 1900s maintained that peace was the cardinal point of the economy and applied themselves, even personally speaking, to support peace initiatives.

War again became, not welcome, but in some cases necessary, in the analysis of Friedrich List in the first half of the 1800s when he faced up to the economic reality of the German states and their need to achieve political unity and reach economic take-off. In this, he criticized the *cosmopolite* economics of Smith and free traders in general, who wished to substitute the presumed universal laws of trade for national and popular sovereignty of politics.[1]

The heterodox free trader Hobson, writing at the beginning of the 1900s during the Second Anglo-Boer War, accused the monopolistic and imperialist tendencies of various entrepreneurial groups supported by the state which favoured war to the detriment of the community. Marxists, as is well known, attributed to capitalism, in its monopolistic and imperialist version, the permanence of armed conflict. The neoclassical, marginalist economists instead took the path of theoretical analysis, setting aside the phenomena that upset the search for a general economic equilibrium. They judged defence to be a problem for the political sphere, independently of all the economic implications that war always fostered.

Some economists and many historians in the interwar period attempted to quantify the material costs of conflict and devastation and to define the particular characteristics of the state economy that had been tried out during the war.

In the period after the Second World War, economists with a mathematical training applied Game theory to the study of military and economic strategy, even if they paid no interest to the causes leading to war, nor in its consequences, thereby making war a new object for study in theoretical economics. They considered war to be a rational activity, albeit part of an exclusively theoretical and mathematical-logical argumentation that could be applied indifferently to the Punic Wars, the World Wars or that against terrorism. They calculated the strategic moves and suggested that deterrence was the fundamental factor in avoiding new devastating conflicts.

The economists of Public Choice too held that the choice of war could be economically rational and thought that it would be opportune to entrust the defence of the state to private bodies, if only for reasons of economic efficiency.

The unexpected proliferation of wars in the closing decades of the 1900s led many politologists, geographers and sociologists, but only a few economists, to take up again the ideas of geopolitics. Some sociologists and economists reread the realism of Hans Morgenthau and Raymond Aron in a neorealist or structural realist fashion. These currents believed they could find the causes of contemporary wars in political factors, which, for geopolitics, were to be sought in questions of territory, while for neorealism the reasons could be found in the situation of chaos reigning in the international relationships between states.

Economic historians, but also some economists, analysed the positive fallout from technological innovation made in the armaments sector on peaceful industry and evaluated the overall economic benefits that derived from war expenditure. Some economists and many pacifists have, on the other hand, calculated the opportunity costs of wars and have denounced the corruption and web of interests that have grown up in all countries between arms-producing companies, the military high command and politicians. In the specific case of the United States during the years of the Cold War, various authors spoke of a military-industrial complex and state capitalism, "Pentagon Capitalism", which dominated the economy and politics of the country. However, generally, it was not professional economists, but politologists, sociologists and historians who formed the body of students which tried to calculate the profits made by the armaments multinationals and contractors.

With the new historical proposal of particular economic circumstances (crisis, market contraction, protectionist policies, etc.), theories which were previously elaborated under analogous circumstances in the past (liberalism-neoliberalism, realism-neorealism) are sometimes adopted again, updated or confuted. Nevertheless, the impression is that economists have paid less interest than other social scientists to the themes raised by warlike conflict over recent decades. The fact remains that the neoclassical, marginalist and monetarist economists, still forming the mainstream in the discipline, at least at an academic level, consider theoretical situations that exclude the hypothesis of war. Exception can be made for the mathematical economists of Game theory and the minority current of Public Choice.

Meanwhile, there are still many wars underway in the world and new conflicts break out in areas limited to a greater or lesser extent, and the news all too frequently reports them. In the period between the end of the Second World War and 2008, according to Monty G. Marshall, Director of the Center of Systemic Peace, 371 armed conflicts have been recorded that have led to an estimated 25,638,850 deaths (cfr. Lemennicier 2010).[2]

More than economists, between the end of the 1800s and the beginning of the 1900s, it has been literary figures and philosophers who have shown greater foresight over the themes of war and peace, even in connection with the economy. They adopted a decidedly pessimistic tone: Burckhardt, Spengler and Nietzsche all predicted that the 1900s would have been a period of great wars because "the spread of western civilization would offer the powers in the field an enormous reward: world domination". The "Spenglerians" opposed the free trade economists when they held that in contemporary society industry was first and foremost at the service of war. Spengler, in particular, on the eve of the First World War, wrote that industrial society was about to unleash one of the greatest wars in history because the concentration of the urban masses, the power of money and the domination of plutocrats and demagogues made a clash inevitable between wills to power masked by changeable ideologies (cfr. Aron 2003, pp. 6–7).

In fact, widespread industrialization and the growth of wealth did not pacify European nations, which, on the contrary, in 1914 entered into the most destructive war for domination ever fought. "Western society, which should have led humanity along the road to peace as its ultimate goal, was instead the birthplace and incubator, but victim too, of the great wars of the twentieth century" (Aron 2003, p. 13).

At the end, in 1918, amidst the general dismay, it was not the economists who made the most important contribution to the analysis of the reasons for what had happened. Kuznets and other economic historians began to evaluate the costs of the conflict for the economy in general and, more often, for individual productive sectors. In 1919, Keynes indicated the dramatic economic consequences that a profoundly mistaken peace treaty would have and, with Einaudi, saw the true solution to conflicts in European unification.

The end of bipolarism, after the Second World War, seemed to have opened the road to the unrivalled domination by the West, to global pacification and to the end of history, but instead has produced global chaos, fear, political and economic closure, the return to small reactionary countries and wars.

While the first half of the 1900s saw two World Wars, both of an "industrial" type, with a technological capacity fought between states, the closing decades of the century have seen the way of waging war change, with the greater strength and technology of the West often being shown to be ineffective, or at least inadequate, when facing new forms of violence.

The end of European colonial empires, followed by the fall of communism, has not brought the hoped-for peace. New freedoms have not "realized the hopes of liberalism" (Black [2004] 2007, p. 85). Warlike conflict, repression and internal violence, often extreme, have marked the end to post-war decolonization, when the former colonies had attempted to form nations, establish boundaries, ensure security and develop their policy of power.

After the fall of the Soviet Union, the United States was able to undertake military intervention in states or regions in the Middle East or Eastern Europe which previously would have gained support from the Soviet Union. Bipolarism is over and the military empires have abandoned the field in favour of technological empires, to use the term coined by Ennio Di Nolfo (2008), that have unleashed further wars. New economic actors have entered the world stage to haunt American predominance, thus outlining a polycentric future. The threat of global terrorism and the fear of aerospatial electronic wars have raised their heads.

The democratic peace of Kantian inspirations has not been reached, wars have not become obsolete, as had been imagined optimistically by John Mueller (1989) and Carl Kaysen (1990) during the crumbling of the Soviet Union. The "post-military society" thought of by Martin Shaw (1991) has not come into being as a society showing no interest in territorial expansion and conscious of the anti-economic nature of war. On the contrary, Bush Junior had returned to the military instrument and the end of the century that was judged as being "short" has seen, even in Europe, the explosion of interethnic conflict that had secondary economic reasons, even if not to be neglected, but even stronger nationalistic, if not tribal, motivations.

Thus, starting in the 1980s and 1990s there has been a strong increase of armed conficts of every kind on a global level. While the First Gulf War, the Falkland/Malvinas War in 1982, the Ethiopian Eritrean War in 1998 and a few others displayed the characteristics of old wars between states, they were followed by prevalently infrastate wars and, starting at the beginning of the present century, the Malacca Strait has seen such an increase in the level of piracy as to make Lloyds of London declare it a war zone in 2005.

Our century has also seen the appearance of non-conventional wars which are totally asymmetric in terms of the forces in the field, in terms of the ends set by the parties in conflict and the very concept of war itself, as considered by the combat troops.

At the same time, state authority in the third world has often seen a loss of the monopoly of violence both internally and at an international level. States are no longer the sole agencies able to intervene in a military way. While ethnic groups have returned to combat each other, the state has, in fact, not infrequently delegated the use of force to private bodies.

The new wars come under many names: there are limited and total wars, wars of high or low intensity, asymmetrical wars, cybernetic wars, preventive wars, "humanitarian" wars and wars without end. There can be many typologies and collective violence, according to the criteria to be taken into consideration. Bertrand Lemennicier, a Professor of Economics in Paris, has suggested, among others, the following classification:

Pre-modern wars, modern wars and postmodern wars
Interstate wars, superstate wars, infrastate wars, non-state wars
Wars of liberation or secession
Local, transnational, global or world wars
Religious, ideological, ethnic or territorial expansion wars
Wars fought by conscripts or mercenaries

To which can be added: psychological or subversive wars, urban violence, revolts, insurrections, civil wars and terrorism.

Then, there are the wars in the poor areas of the world, African wars and the many forgotten wars. Not only economists, but even more so, journalists, politologists and missionaries recalled that the "African World War", that began as a civil war in the Congo, was based on the mineral wealth of the country. The Colombonian and the Salesian missionaries are much more careful than economists in seeking the economic causes of contemporary wars and in evaluating the consequences for the population, economic growth and the area involved.[3] It is the Pope himself who constantly recalls the war underway and speaks of "a third world war fought in pieces".

The Gulf Wars have ensured major profits not only for arms producers, but also for the oil magnates, Bush's grand electors, who defined the conflicts he unleashed and potential future ones as preventive wars, destined to become continuous wars.

Thus, after the positivist idea of "the end to war", after a century and a half, we have come to think in terms of "war without end".

"This nation is peaceful, but fierce when stirred to anger. This conflict was begun on the timing and terms of others. It will end in a way, and at an hour, of our choosing [...] It may never come to an end, at least during our lifetime". Bush and Cheney stated this in 2001, after the attack on the Twin Towers.[4]

In 2006, the Quadrennial Defense Review Report, an official document of the American Department of Defense on strategic and military priorities of the United States, warned that: "Currently, the struggle is centered in Iraq and Afghanistan, but we will need to be prepared and arranged to successfully defend our Nation and its interests around the globe for years to come" (https://archive.defense.gov/pubs/pdfs/QDR20060203.pdf, p. V).

Recent decades have seen progress in economic and financial integration, known as globalization, continue worldwide, while at the same time "political space" had disintegrated still further. New, anachronistic minnow states have sprung up. This has led to the return to fashion of geopolitical theories to explain conflicts.

Often states can no longer control strategic resources, while multinationals, often with a capital rating higher than the GDP of many countries, have managed to condition and define general economic development. Then again, the new conflicts unleashed by the United States, leaving aside official declarations, have an often open link, if not the only one, to the problem of energy supply. This has led to reconsidering the old problem of "shortage" as a cause of war: oil shortage, raw materials shortage, water shortage.

Some, instead, seek other motivations, beyond economic and political ones, for the aggressiveness of the United States in recent decades and often find them in the influence on American thought of the 1800s concept of "manifest destiny", a variation on the theme of "destined by God", which English puritans had imported to the United States.

According to this, "it is no longer just the borders of the Kingdom of God that have to be pushed back, but also those of democracy" (Ricard 1986, p. 61). Bush Junior too taught about destroying "monsters" in the name of the war against evil.

Reginald Stuart and Huntington have stressed the dichotomy in the American mentality due to the simultaneous acceptance present in rational liberal thought and the tendency to transform every war into a crusade in defence of universal principles. Kissinger himself stressed that, while the realist Theodore Roosevelt was convinced that the preeminent role of the United States was dictated by national interest and the need to balance forces on a world scale, Woodrow Wilson, instead, had a messianic conception of the United States which, as it was incarnated in virtue, had to intervene not so much in order to define international equilibria, as to spread its principles throughout the world (Kissinger 1994; cfr. Hassner and Vaïsse 2003, pp. 24–26).

But, at the end of the day, economically speaking, who gains from a war which drags on possibly to infinity? Are there, today, any limits to war?

Two Chinese Colonels, Qiao Liang and Wang Xiangsui (2001, p. 25), have spoken of present-day conflicts as wars without limits, because juridical barriers in them have broken down. Those barriers laboriously set up over time with *jus in bello,* and consequently, the restrictions and taboos that distinguished military personnel from civilians and the armed from the unarmed have been surpassed. Today, non-military wars are fought and military operations other than war are carried out. The ratio between civil and military victims has been overturned; there is no longer a limit to the means used, from machetes to missiles, from hijacked civil flights to computer viruses, not forgetting manipulation of the media. There are no longer any geographical limits as battlefields have no barriers and war can be waged anywhere, even in the heart of the United States.

Often wars are linked and the fighters can be deployed in different geographical areas while maintaining the same communication channels and logistic supply lines. There are no longer time limits to fighting, as "many wars do not end, but remain covered up, flare up again with combat in neighbouring zones, changing form and protagonists. National type conflicts are substituted for by religious motivations". So it is often difficult to see if a conflict has ended or not: "some present day wars resemble underground rivers which can re-emerge after a while in the same place or in nearby places" (Desiderio 2008, pp. 15, 140).

Publications on war from various points of view—military, philosophical, juridical, religious, biological, psychoanalytical, social or strategic—are seemingly endless. The analysis carried out by economists too, even if it is perhaps not the most fully developed, is difficult to handle in just one study. Any summary, including the present one, turns out to be partial and open to legitimate criticism over omissions and incompleteness. In this study, war is considered as armed conflict between states. It therefore does not take into consideration civil wars, insurrections, coups d'état, violence and internal wars, which, in any case, have not been the object of particular reflection on the part of economists, if we exclude mercantilists and the students of Public Choice.

The choice of authors has not been based on their fame, but rather on that of the originality of thought expressed on the theme of war. The aim of the work is to seek to understand how far the position of economists on war is dictated by the situation of war in their time and by the vision they have of the subject they were studying: economics is an exact science or a social science? Do a-temporal, universally valid laws guide the chimera of *homo œconomicus* in the choices made and the management of the use of violence as well?

## Notes

1. Perhaps, it is worth noting how international law in the same period was developing a parallel critique with an accusation of the English undermining the bases of *jus publicum europaeum.*
2. Maurizio Simoncelli (2005, p. 27) calculated according to "prudent estimates" about 23 million dead (mainly women, old people and children) between 1946 and 2000. In the same period, according to his calculations, conflicts on a world scale were "more than 150". Perhaps, Simoncelli included in the count only what were conventionally called "wars", that is, conflicts that cause over 1000 deaths. Naturally, all estimates have a wide degree of approximation.

3. Besides the news provided by "Nigrizia", the monthly publication of the Padri Colomboniani and by the regular publications of the Salesians, in 2005 Caritas Italiana published a well-documented report on "Conflitti dimenticati, guerre infinite, terrorismo internazionale" [Forgotten conflicts, endless wars, international terrorism] (P. Beccegato, W. Nanni, F. Strazzari).
4. G. W. Bush (2001); R. Cheney, declaration on 19 October 2001 "Washington Post", 21 October 2001. See also Tertrais (2004, p. 5).

## References

Aron, Raymond. [1957] 2003. "Guerre e società industriale" in id. *Il ventesimo secolo*. Bologna: il Mulino.

Beccegato, Paolo, Nanni Walter, and Francesco Strazzari, eds. 2005. *Guerre alla finestra*. Bologna: il Mulino.

Black, Jeremy. [2004] 2007. *War Since 1945*. London: Reaktion Books Ltd.

Bush, Georges W. 2001. *President's Remarks at National Day of Prayer and Remembrance*, 14 September.

Desiderio, Alfonso. 2008. *Guerre del 21° secolo*. Firenze: Giunti.

Di Nolfo, Ennio. 2008. *Dagli imperi militari agli imperi tecnologici. La politica internazionale dal XX secolo a oggi*. Bari: Laterza.

Hassner, Pierre, and Justin Vaïsse. 2003. *Washington e il mondo. I dilemmi di una superpotenza*. Bologna: il Mulino.

Kaysen, Carl. 1990. "Is War Obsolete? A Review Essay." *International Security* XVI (3): 731–745.

Kissinger, Henry. 1994. *Diplomacy*. New York: Simon & Schuster.

Lemennicier, Bertrand. 2010. *De la guerre*. Paris: Panthéon-Assas, Université Paris 2, Cours second semester 2010.

Liang, Qiao, and Wang Xiangsui. 2001. *Guerra senza limiti. L'arte della Guerra asimmetrica fra terrorismo e globalizzazione*. Gorizia: Libreria Editrice Goriziana.

Mueller, John. 1989. *Retreat from Doomsday. The Obsolescence of Major War*. New York: Basic Books Inc. Publishers.

Ricard, Serge. 1986. *Theodore Roosevelt et la justification de l'impérialisme*. Aix-en-Provence: Université de Provence.

Shaw, Martin. 1991. *Post-Military Society: Militarism, Demilitarization and War at the End of the Twentieth Century.* Cambridge: Polity Press.
Simoncelli, Maurizio, ed. 2005. *Le guerre del silenzio. Alla scoperta dei conflitti e delle crisi del XXI secolo.* Roma: Archivio Disarmo.
Tertrais, Bruno. 2004. *La guerre sans fin. L'Amerique dans l'engrenage.* Paris: Editions du Seuil.

# 2

# War and Economic Activity

## International Trade and War

Political power and wealth in feudal society were identified with landed property and land was conquered by violence on the battlefield. Nevertheless, war was not studied in relation to economics, which in any case was not yet an independent science.

The first economists to consider the relation between war and economy were the mercantilists, starting in the second half of the sixteenth century, the same period in which nation states were born and shortly after the birth of modern political thought, thanks to Machiavelli, who established, among other things, the necessity and opportunity for the state to use violence.

The reflections of the mercantilists did not give rise to an organized and structured doctrine, but instead led to practical considerations and suggestions offered to the governments of that time. Mercantilism thus became both the policy and the economic theory of the rising national states.

According to the mercantilists, the economy must set its objective as that of enriching the state, identified with the sovereign, to offer stability


R. Allio, *War in Economic Theories over Time*,
Palgrave Studies in the History of Economic Thought,
https://doi.org/10.1007/978-3-030-39617-6_2

and support his power. But seeing that the productive techniques of the period did not allow for a significant level of enrichment in the short term through improving agriculture or manufacture, it was found to be easier to try to seize the wealth of other states, thereby reaching the twin goals of reinforcing the power of one's own country and destroying that of the others. This meant that war had a fundamental role and that there was a close link between economics, politics and warlike aggression.

Economic theory, having been called on to resolve the problems of the state, turned out to be subordinated to the needs of politics and was warlike because it reflected the reality of those years, when the newborn states fought among themselves to establish their own frontiers, while at the same time fighting against the residues of feudal privilege and the power of the church, but also for possession of the wealth discovered in the New World. War, in turn, was seen as an economic resource because it gave the victor the right to dispose of the property and land that belonged to the defeated state. In deciding on war or peace, like in deciding any other question, the sovereign simply based his decision on the calculation of his own interests, or, stated more elegantly, for "reasons of state".

In any case, the only limit posed by the mercantilists in reaching the decision to unleash a war was that of not having a sufficient degree of probability of victory, as William Temple stated explicitly in an essay in 1673 on Irish trade. Victory in war required an adequate army, while Francis Bacon held that the most useful element forming the greatness of the state was "a race of military men" ([1625] 1944, p. 93). Bacon compared the human body to the political one and stated that both required physical exercise to stay in form, so while a just and honourable war was the best physical exercise for the kingdom or state, a civil war was like a fever. However, the concept of a just war was not posed by Bacon in relation to the concept of justice. According to him, in fact, states must establish laws and customs that can offer "just" opportunities, or at least plausible pretexts, to unleash conflicts (Bacon [1625] 1944, pp. 100 ff.). Machiavelli too held that every necessary war was a just war, and every war useful in enriching the state was necessary.

The mercantilists did not agree among themselves over whether it was better to fight a war in the home country or abroad. Many of them thought that it was better to fight the battles abroad so as not to cause damage to property and activities of the home country, but others, including Charles Davenant, held the opposite opinion based on the consideration that wars conducted abroad implied taking money out of the state. Many stressed that wars were becoming even more burdensome above all with the establishment of large permanent mercenary armies, the renewal of both terrestrial and maritime military techniques and the employment of massive armed galleons.

Apart from its economic function, war was considered as a fundamental institution because it served to maintain the honour and the virtue of the nation and drive away the spectre of civil war, the real ruin of states, according to the mercantilists. Jean Bodin exemplified this concept in the chapter entitled *Guerres des ennemis necessaires pour entretenir les états populaires* as follows: "This was the main reason which led Scipio the Young to hold off as long as possible the razing of Carthage, because, as he wisely considered, the Roman people, which was warrior like, and warlike, would no longer have had enemies and would have inevitably slipped into civil wars" ([1576] 1986, vol. IV, p. 35: author's translation).[1]

The idea that war against an external enemy offers an antidote to internal revolts crops up frequently in the mercantilists' writings. The same was true of the idea that it is not war, if the state is strong, but rebellions, which are the true cause of the fall of empires and republics. Preventive wars are therefore suggested to avoid civil revolts, even creating external enemies for the occasion, if none such already exist. A thought also expressed by Machiavelli.

For Botero too, wars caused by whatever reason, "whether it be waged to secure frontiers, to add territory to the dominions, or simply to win glory and riches, to protect allies or assist friendly powers, or to defend religions and the worship of God", have always been a great and general "entertainment" for the whole population, because

> everyone who is able is ready to play his part either in council or in action, and all discontent is vented on the common enemy. The rest of the people either follow the camp to bring supplies and to perform similar necessary services, or remain at home to offer prayers or vows to God for

> ultimate victory, or at least are so stirred by expectation and by news of the progress of the war that there is no place for thoughts of revolt in their minds. (Botero [1589] 1956, pp. 76–77)

Similar reflections can be found in the *Traité de l'économie politique* by Montchrestien and the *Traité de la nature* by Boisguilbert (1707).

Another consideration common among mercantilists as a whole regards the need for states to be prepared for war as a fully legitimate and historically recurrent activity. This means that substantial funds have to be set aside to finance armament, because, according to the formula expressed by Marshal de Saxe[2] and since then often repeated, war needs three things: firstly, money, secondly, money and thirdly, money.

Nevertheless, as Tomas Mun stressed in this case, if money is the backbone of war, in order to be useful, it has to be invested correctly to organize the army beforehand and to see to the necessary stores and munitions (Mun [1664] 1928). Colbert recalled instead the need for the state to see to navigation and war industries in preparation for conflicts.

Besides the appropriation achieved through war, the wealth of the state can also be increased by a flourishing and profitable foreign trade, thanks to which it would be possible to increase the availability of precious metals that served, among other things, to pay mercenaries and so conduct the war. In turn, war could help trade when it is aggressive and aimed at the conquest of colonies. There was therefore a synergy between trade and war, with trade becoming an "affair of state". This was all the more so with the emergence of the two great maritime powers, Holland and Great Britain, which eclipsed the old Renaissance trade centres (cfr. Hont 2005, p. 23).

It is therefore no coincidence that the first analyses by the mercantilists of the relationship between economic activity and war all refer to international trade. The surplus to the trade balance is obtained by exporting as much as possible and holding back imports as much as possible, even at the cost of organizing a domestic regime of an autarchic type.

Government activity aimed to organize and control production and exchange with the scope of assuring international predominance by the state. It was vital to be richer and more powerful than enemies, so foreign trade was therefore seen as an effective means of national enrichment, part of the holy egoism of the state. Foreign trade not only supplied at the start the money needed to pay the army, but the merchant fleet and civilian sailors too could be used for military ends, if so required.

Both Botero (1589) and Montchrestien (1615) held that trade was the simplest and most widespread way to become wealthy at others' expense. The state economy was consolidated by a surplus trade balance and any such military intervention that trade could lead to would not be bad; in fact, it could be used to avoid civil wars, the bugbear of mercantilists. Colonial wars had the same useful effect.

If war of aggression was judged legitimate by mercantilists, both French and English, trade, even during peacetime, was considered offensive, because it tends to improve the economy of one state by wilfully damaging that of another at the same time.

"Trade today is one of the main parts of politics" wrote Huet, still at the beginning of the eighteenth century (1712, Preface, p. 19), and its role is offensive as it intends to contribute to realize the economic domination of the state, bringing it wealth to the detriment to other countries. The economic interests of various nations were seen as incompatible: a country could only grow wealthier at the expense of others. Trade, like war, was zero-sum game. This policy, when concretely realized, fanned conflicts in Europe over the establishment of the frontiers of national states and upheld colonial conquests.

The wish of mercantilists to increase the wealth of a state did not have as its aim, not even a secondary aim, the improvement of the living standards of the subjects.[3] Mercantilists largely agreed with Machiavelli when he thought that, in a well-ordered state, the subjects are poor and the prince rich.

Jean Bodin, a theorist of absolutism and important spokesman of mercantilism, took a different position on foreign trade. In his *Réponse à M. de Malestroit* (1568), Bodin was a forerunner of laissez-faire thought when he stated that free trade was the best means to guarantee peace and friendship among peoples because in free trade neither of the parties suffers damage. He also added that peace is a duty imposed on sovereigns by providence. Nevertheless, just eight years later, Bodin offered a completely different opinion, returning to that of the mercantilists. In *Les six livres de la République* (1576), it is now war that is seen as providential and the source of wealth for the state. Here, Bodin holds that the struggle is not to obtain peace, but to conquer the property and land of the defeated states. The aim of war is not to reach an agreement, with the consequent pacification, but rather to enrich the state.

Two centuries later, Pietro Verri again demonstrated the bellicose intent in the trade of his time, considering "any advantage to a nation in trade leads to the damage of another nation, the study of trade which today expands is a real war which is silently conducted by various peoples in Europe" (Verri [1760] 1804, p. 335).

Nevertheless, starting at the end of the seventeenth century, some economists, both French and English, had begun to reflect on the negative nature of aggression in exchange and to propose a different version of international trade, as Jean Bodin had already done in the mid-sixteenth century, but only to withdraw his remarks later. Gradually, the activity of exchange came to be understood as civilizing and pacifying, at least by some thinkers who followed the ideas of the physiocrats and laissez-faire economists. Free trade and brotherhood among nations from then on began to be considered future goals, but not without a theoretical to and fro and some incongruence. Abbot Galiani is a famous case in point. He criticized the mercantilists over their monetary theory and the physiocrats over the free market in cereals.

Jacques Savary, in his book *Le parfait négociant* (1675), spoke for the first time about the "sweetness" of trade, but the popularization of this fresh characteristic attributed to trade took place only much later with the publication of *L'esprit des lois* (1748) where Montesquieu holds that trade sweetens barbaric habits, so that wherever trade takes place, customs are sweet. The interest of who intends to sell and who needs to buy meets in exchange, the interest is mutual and, consequently, the "natural" effect of trade is to bring peace and civilization. In fact, the operation of exchange not only tends to reduce wars between nations, but also has a civilizing effect.

Among the writings of the first economists selected in 1856 by McCulloch, we can find an anonymous (William Petyt?) pamphlet dated 1680 entitled *Britannia Languens, or a Discourse of Trade*, which contains the following judgement on the effects of war:

> I shall admit that if a Nation can be *Victorious in War*, and can plunder the Conquered, some treasures may happen to be imported in this way; but certainly those who consider it, will rather desire to be enriched by Trade than by War, since in the Course of Trade, far mightier Treasures may be gotten with *Peace, Innocence, Security and Happiness to the People*, who cannot be Victorious in war, without Bloodsheds, Rapines, Violences, and Perpetrations of all kinds. (McCulloch [1856] 1995, vol. 1, p. 291)

Sir Dudley North, an English merchant, politician and economist, whose *Discourses upon Trade* was published anonymously in 1691 and was also included in McCulloch's *Selected Collection* of 1856, also expressed himself in favour of free trade. A similar position was adopted years later by Jacob Vanderlint, a merchant of Dutch origin who had lived in Britain, in a text called *Money Answers All Things*, published in 1734, which also deals with trade.

Reflections of the same type can be found in France in a period when, nevertheless, England was seen as the monopolist in foreign trade.

Jean-François Melon (1675–1738), the financial counsellor to the Regent (Philippe d'Orléans) in the 1720s, later the secretary to John Law and an influential member of the cultural circles; and the Marquess d'Argenson (1694–1757), Secretary of State for Foreign Affairs under Louis XV, a friend and correspondent of Voltaire, surveyed physiocratic thought on the question of exchange and questioned in their writings the idea of the economic usefulness of aggressive trade and conquest, holding that it was more advantageous for all concerned to follow exchange which offered benefits to both sides.

Melon was convinced that trade by itself was a pacific activity, even if competitive. Nevertheless, while it was true in the past that economic needs of war led to the expansion of trade, in future the needs of trade could cause wars. The solution for states did not lie, however, in substituting wars for power with wars for trade, but rather in seeking to realize an international economic balance. If this balance were not to be found a coalition attack against the country holding the trade monopoly, it would have had the character of a just war and would have been approved by the Laws of Nations.

In order to avoid conflicts, all European nations would have to go further than just hindering the constitution of trade monopolies by becoming self-sufficient in food production, seeing that the food supply was often the cause of war (Melon 1734, ch. 1, pp. 1–12). The violence of war was not in any case the suitable means to correct economic unbalances, so Melon encouraged France to compete economically and not just in a military way with Great Britain, hoping thereby that the latter would direct its economic potential to a pacific idea of trade rather than to the bellicose wish to become a conquering nation.

The Marquis d'Argenson (1764), while denouncing the anti-French feeling and English "jealousy" in its own international trade, maintained that each country was all the more prosperous, the more prosperous were its neighbours, therefore condemning the protectionist system and customs barriers, be they English or French, which, by attempting to reduce imports, ended up by damaging exports through reprisals.

In the second half of the eighteenth century, along with the new way of seeing trade, a different way of judging other productive activities, above all agriculture, was gaining ground. In France, a new current of thought, physiocracy, considered agricultural activity as the sole source of wealth in a situation of peace and international harmony. This thought developed in connection with the improvement of agricultural techniques with the consequent increase in the productivity of the land, on the one hand, and the spread of the ideas of the Enlightenment, which stated that the natural order also rules production, on the other.

Mercier de La Rivière, a physiocrat, was convinced that a common interest tied all nations together and, in his book: *L'ordre naturel et essential des sociétés politiques* (1767), upheld the reasons for free trade, stressing that it was absurd to think only of exporting without ever importing anything. After considering the relations that must "naturally" or "necessarily" operate between nations, he concluded that populations that maintain trade relations, despite the distance separating them, succeed in constituting a pacific community. In fact, according to him, every nation is just a province of the great Kingdom of Nature.

However, this order was not observed in the reality of his own time, and European nations continued to base their foreign policy on the wish to enrich themselves or expand at the costs of others, by seeking as in the past to gain great profits through aggressive trade. However, according to de La Rivière, it was "physically impossible" for a policy that ran against the interests of other nations not to turn them into enemies, provoking war and general damage. This "false policy" had to pay a high price for the presumed advantages which "because of the wars they determine, compromise the security of a state. And which, if a deeper examination is made, do not just disappear, but instead are converted into privation, into real losses for the nations and their sovereigns who have been seduced by these illusory advantages" (de La Rivière [1767] 1910, p. 244).

Mercier de La Rivière, however, remained optimistic: despite the different and negative political approach of the period, the brotherhood of nations and the creation of a confederation of all the European powers did not seem to be passing utopian fancies to him. This was precisely because a general confederation represented the "natural" political condition of Europe, so it would have to be realized sooner or later.

More generally speaking, according to de La Rivière, peace can be achieved just by observing natural laws; there is no need to set up international organizations. The state should not intervene in the economic field, as the mercantilists thought, but instead here too should let spontaneous natural order act freely.

Physiocratic thought did not attribute a particular value to the accumulation of precious metals and trade, which it considered in itself sterile. Consequently, an economy based mainly on trading, as proposed by the mercantilists, appeared to be inevitably the source of damaging rivalries and international conflicts for the conquest of markets, above all those supplying raw materials at a low cost. On the contrary, foreign trade, just like domestic trade, must be free and must not seek to damage the other party, but, instead, must be based on the reciprocal advantage of nations that exchange goods at the same value, thereby becoming a positive factor in convergence and pacification.

Pacific laissez-faire in international relations thus ran counter to bellicose mercantilism. Whereas the mercantilists sought to increase wealth and power of their own state, taking it away from other states through war, the laissez-faire school, which considered human labour, sought instead to increase production and the private well-being of citizens through peace. They believed that wealth could and should be increased by improving domestic production and consequently thought that nations had a common interest in the free and pacific trading of their respective surpluses. As opposed to mercantilist thought, the physiocrats and classical economists were convinced that the mechanism of free trade would have led to an equitable division of benefits between the parties and, consequently, to general prosperity and thus peace.

Laissez-faire economics developed theoretically without being subordinated to politics, unlike the case of mercantilism. Its aim of enriching citizens led at the same time to increasing the wealth of nations, this being simply the sum of the wealth of citizens. The liberal states, unlike the aristocracies, would no longer have aspired to economic supremacy to the detriment of competitors, because both the physiocrats and the classical economists saw a common interest linking men and nations.

The ideas of Mercier de La Rivière were equally held by Quesnay and the physiocrats in general: in the economy *laissez-faire, laissez-passer* sufficed and there was no need for state intervention. International trade should be completely free because it is a factor in prosperity and peace among nations.

The wealth and power of a state cannot be founded on the ruins of others because the natural order makes the interest of all a common one. Wars caused by economic reasons arise from the fact that nations are made to be the interpreters of the interests of certain groups and guarantee their exclusive privileges to the detriment of general interest.

> All wars and all the limitations placed on trade can only have the object of a monopoly [...], which is always a curse on those nations which do not distinguish their interest from those of their traders and which are ruined by conducting wars to assure an exclusive privilege to the national agents of their trade at prejudice to themselves. (Quesnay [1759] 1888, pp. 239–240, 461, 467–468, 489–490)

Quesnay stated that he was convinced that wars rarely break out in a well-governed state.

Production and commerce increase national wealth, while closure and autarchy are aims that suit no one. On the contrary, trade is to be favoured even between rival or hostile states. Quesnay in fact warns that trading nations must pay greater attention to maintain and, when possible, to expand their trade instead of seeking to hinder directly that of others, because trade between nations is mutually supported by the wealth of buyers and sellers.

Giuseppe Palmieri, in Italy, the author of, among others, an essay on the art of war (1761) maintained in his *Riflessioni sulla pubblica felicità relativamente al Regno di Napoli* (1788) that war "is the greatest of evils", not only because it means a greater outlay by society, but also because it is also the activity "which is most against the well-being of citizens, apart from those people that live off it and the infamous and illustrious great thieves both on sea and on the land" (Palmieri [1788] 1805, p. 334). In the past, for example in the Roman period, war could have brought economic advantages, but in the eighteenth century this was no longer the case and a good government should do its best to avoid it. In particular, nothing in Palmieri's eyes appeared "stranger, more absurd and more contradictory than war undertaken by trading nations for reasons of trade" (Palmieri [1788] 1805, p. 335) because war destroys just those trading activities, it is hoped to undertake or increase. Another evil is a great standing army, because "the state thereby runs up debts for an overwhelming army causing a lack of revenue which many people could produce if employed in other trades" (Palmieri [1788] 1805, p. 336).

David Hume too was against obstacles, barriers and duties hindering trade and dedicated a chapter of his *Political Essays* to the *Jealousy of Trade*[4] to condemn the position of countries, which are alarmed when they see the economic growth of their neighbours, considering them as always their enemies. Hume held that the increase in the wealth and trade of any nation, far from damaging the interest of the others, almost always contributed to increase their opulence, thanks to the increase in trade and the new wealth created. In fact, a country cannot increase its trade and industry if "all the surrounding states are buried in ignorance, sloth and barbarism" (Hume [1752] 2000, p. 63). As a British subject, Hume strongly wished for growth in trade in Germany, Spain, Italy and even in France itself and stated that he was certain that Great Britain and all the other countries would see their prosperity grow if their sovereigns and ministers opened to sentiments of reciprocal benevolence (Hume [1752] 2000, p. 66).

Hont noted that jealousy in trade was mentioned when success in international trade was adopted as an element of survival or the political and military development of the nation, that is, when economics has become politics. Nevertheless, as Hume stresses, war and commerce follow two different logical paths: war sees the victor and the vanquished, but trade is based, instead, on reciprocity and a mutual benefit. Jealousy of trade therefore describes a process of corruption because it implies that the logic of trade is forced into the logic of war (cfr. Hont 2005, pp. 5–6).

Smith saw progress in the development of industry rather than agriculture, as was the case for the physiocrats, but was equally convinced that individual interest coincided with that of the nation and that this, in turn, coincided with international interest. This allowed for the development of a global market, even with the continued existence of nation states. From this point of view, free trade is an advantage for all nations. It favours peace and creates the cosmopolitan trader, the citizen of the world. However, in reality, international trade was still often dominated by the East India Company's wish to establish a monopoly, fed by the rapacity of traders, and consequently leading to results diametrically opposed to its potential. The wish to enrich oneself at the cost of others survived and provoked, among other things, wars, which damaged both populations and their economies.

Jeremy Bentham was completely opposed to war and for a whole series of reasons: because it was a typical characteristic of barbaric peoples and was anachronistic in the nineteenth century, because it destroys human lives, because it damages the economy, because it is a means through which governors increase their power, because it encourages patronage and because it leads to tax rises. Bentham held that the harmony of interests, that existed not only between individuals, but also between states, if given free rein, would produce well-being for all. Free trade, for example, would allow earnings higher than could be realized through policies of territorial expansion, and so too peace allows greater production, guarantees freedom of trade and is much more profitable than war.

If war is irrational, but is conducted all the same, and the reasons are to be found in the behaviour of autocratic governments to be corrected either by the dissuasion on the part of public opinion or by the work of an international organization. Economic progress is nearing the moment in which reason will prevail in international economic choices. The transition from autocratic and despotic forms of government to representative democratic forms will lead states to pursue the interest of citizens and not those of the sovereign or minority social groups, so opening the path to peace. As regards colonies, Bentham held that they can only be useful for populating, while they are not so if they are conquered with the aim of a presumed enrichment of the motherland (Bentham 1873, pp. 852–858).

Bentham nevertheless believed that free trade alone was not sufficient to realize peace, the indispensable condition for general well-being. Neither the natural order nor free or "sweet" trade were enough to guarantee the general harmonization of interests. It was therefore necessary to set up an international organization of a political type with powers of control and management, with a common tribunal of justice to judge controversies between nations. This tribunal, as Bentham saw it, would not have coercive powers, but seeing that its sittings would be open to the public, being in a majority in favour of peace, it would act as an instrument for sanctions against those states ignoring the directives of the tribunal.

For McCulloch too, foreign trade, if free and not monopolistic and is founded on reciprocal interest, achieves the satisfaction of needs and mutual advantage. Military conflict, instead, means huge expenditure to be borne by national producers through taxation. Seeing that wealth is the source of power and free trade is the source of wealth, a state that seeks to become powerful must eliminate any check on international trade.

McCulloch considered political economy to be a pacific science. If a war breaks out for economic reasons, the premises on which these reasons are based must be incorrect. The mercantilist choices in favour of protectionism, war and colonialism did in fact slow development, depopulate and ruin countries wishing to gain wealth.

In his introduction to the 1872 edition of *The Wealth of Nations*, by reconstructing the contribution of economists from the mercantilists to Ricardo, McCulloch held that it was the task of economic science to demonstrate that men have every interest in avoiding war and putting into force the principles of free trade (1872, pp. xv–xvi).

Manchesterism too proposed free trade as a factor in international peace and prosperity and, according to Cobden, economic science itself had to prevent international conflicts because war is a costly luxury and a blight on the nation. International trade must be free and does not require the support of armies and fleets as economic competitiveness is its invincible arm. Colonization too should not be protected by arms. If it is necessary to keep control of colonies by violence, it is better to grant them autonomy. Free trade leads to economic prosperity and is the only way to reach permanent peace and well-being for the population. States must abandon the policy of territorial conquest and reduce their armaments. Arms and warships cannot really protect or extend trade and the cost of maintaining them weighs down and hinders the development of manufacture by burdening it with taxes.

The widely used English saying "Ships, Colonies and Commerce" used for toasts at the Pitt Club annual meeting, a saying that could be translated concretely as "a war fleet for the conquest of colonies and to monopolize their trade", had to be abandoned because it was outdated; it should have been replaced by a single word, "cheapness", that is, competitiveness, that of English manufactures that would conquer markets on their own and peacefully (Cobden 1878, p. 125).

Cobden recalled that public expenditure, and therefore taxes, can be reduced by avoiding war and rearmament. In this way, growth and the material well-being of the people are promoted, without putting national security at risk, because the reduction of the apparatus for offence reduces international tensions along with fears, hatred and suspicion among nations. It is a common error to calculate the power of a state on the basis of its military forces, both land and sea, whose exorbitant cost is, on the contrary, a frequent cause of the poverty of the population (Cobden 1878, p. 59).

Arbitration is more rational, more just and more humane than the use of arms in international controversies. It should therefore take first place in realizing peace among nations, while also reducing armaments, engaging in non-interference in the internal affairs of other nations and cutting off credit to warlike governments (Cobden 1870, 1878).

Cobden also defended his ideas on a free market, the reduction of military expenditure and renouncing colonial adventures in the House of Commons. He proposed abandoning the colonial empire during the campaign he led against the Corn Laws, but his idea did not take hold and in 1849 he complained that "the middle class is stuck to the colonial system no less than the aristocracy and not even the workers are more intelligent" (quoted in Sombart [1916], 1927, p. 62).

Dupont de Nemours too overturned the mercantilist conception, saying that peoples carry on bloody wars for ridiculous reasons and no one wants to admit that a war to encourage trade is just a "barbaric extravagance" directly opposed to its objective; one cannot oppose the trade of neighbouring countries without damaging one's own as opposing the sales of enemies means reducing their purchasing power, thus taking away from them means to acquire products which it would be useful or necessary to sell to them (Dupont de Nemours 1769, p. 55).

In the Great Britain of the 1770s, Josiah Tucker, a churchman and economist, while continuing to hold certain positions that were still mercantilist over some issues regarding foreign trade, came out in favour of freedom and against wars unleashed in the name of economic interest and denounced the colonial system. In consideration of the reality of his country at the time, he stressed how warlike conquest of territories in order to further trade was an absurdity since the ability to produce at low cost and competitiveness encourage trade, not colonization. He recalled that landowners, manufacturers and the vast majority of the population are interested in peace, because in wartime the cost of imports rises markedly, exports fall, consumer goods become scarce, their quality worsens and their price increases. Certainly, war can enrich some merchants and offer promotions to army and fleet officers, but their interest conflicts with the general interest of the country, because neither princes nor the people can gain an advantage from wars, even the most successful ones. Trade develops freely in countries where goods are best produced and at a low price, rather than in those that aim to conquer new territory. It would be more opportune to promote mutually advantageous trade, even if it is difficult to convince those who are against the common good to practice it than to prevail by the use of arms.

Whenever technology and economic means no longer allowed England to maintain its industrial primacy, Tucker did not rule out the possibility of taking violent action, even if he hoped that, with the passing of time, warlike behaviour would change. "Possibly, at last, the tide may turn" and so future generations will see "the present madness of going to war for sake of trade, riches, or dominion, with the same eye of astonishment and pity, that we do the madness of our forefathers in fighting under the banner of peaceful Cross of recover the holy land" (Tucker 1763, p. 58).

Lorenzo Infantino picked up from Tucker's considerations to maintain that, apart from the different type of economic organization (aggressive and war based as proposed by the mercantilists, laissez-faire and free trade for the liberals), there was a major continuity between the two schools. In each case, they both boiled down to seeking to dominate in the end, even if this goal was pursued by the classical economists with non-warlike means, but through private economic entrepreneurship, that is without employing the mercantilists' state apparatus. The technological superiority gained by England in the industrial revolution allowed the country to sell its products almost anywhere on a market much greater than just the colonial one, so saving on the cost of controlling and administering the colonized territories.

Infantino starts from an article by John Gallagher and Ronald Robinson (1953) on English *Imperialism of Free Trade* in the Victorian period[5] to underline how the anti-imperialist formula of the classical economists "trade and not government" was rewritten as "trade with informal control where possible" and "trade with government (formal domination) where necessary". The freeing of international trade would not have guaranteed the elimination of war. "The state stood back for the moment" as long as the new means of production succeeded in maintaining primacy, "remaining, however, ready to intervene whenever this domination could no longer be reached 'technologically' and 'economically'. Military force was therefore not suppressed, but transformed from the *prima ratio* into the *ultima ratio*" (Infantino 1980, p. 17).

If we return to the eighteenth-century economists, above all Antonio Genovesi, it should be stressed that summarizing his thought on the relationship between trade and war sometimes leads to some difficulties. Antonio Genovesi held, in fact, that trade, if "well conducted and well governed" produces "many nice effects", above all increasing the wealth and power of the nation. Furthermore, trade was a factor in civilization and can "lead trading nations to peace", uniting them through reciprocal interests which can only be expressed in a state of generalized peace. Consequently, "war and trade are as diametrically opposed as the storm rough and the calm, so that where trade is preferred, it is impossible to go to war [...]" and so far the thought flows, but Genovesi immediately adds "[...] if not to support trade". Therefore, trade leads to peace and only during peace can the common interests of nations be expressed, but war can break out for trade, obviously a just war (Genovesi [1765] 1824, pp. 151–152). However, later on, Genovesi explains that by support for trade he does not mean helping privileges or monopolies, but rather the defence of trading activity so that "it does not find impediments in its way" and so is respected by other nations. This protection can be offered by treaties or the fleet, consequently, "trading nations" must maintain on the seas "good armed vessels to gain the respect of those whose avidity for prey has made them forget the rights of nature". Of course, the way of treaties is "more humane", but "the best protection is always armed force" (Genovesi [1765] 1824, pp. 117–118).

In any case, the spirit of trade is not that of conquest of peoples and lands, typical of barbarians, but, on the contrary, free trade pacifically conquers wealth. Wars damage or hinder trade and "the true power of a nation can be seen in its ability to repel an unjust war with *forza ed arte*, or to be able to wage a just one". Able to in the sense of having the economic means required of "troops, victuals and mechanical arts". This guarantee of availability, according to Genovesi, is provided by a fundamental conduct of flourishing trade.

Exchanges, meetings and the comparisons, which trade feeds, favour the development of science. Trade sharpens the mind. Said so, it is also true that "not infrequently the jealousy of gain and the command of the seas, arms nations and drives them to war, but the interest of trade disarms them in a short time" (Genovesi [1765] 1824, pp. 144, 153).

Genovesi's thought about colonialization is equally complex. The colonies have become a necessity "for a respective and not an absolute reason". Some territories in South America and Africa were "occupied by force" by European countries that had previously monopolized their trade. Consequently, other European nations then saw the need to intervene to gain zones for privileged trade.

Colonies presented advantages as they favoured the development of the fleet and the growth of trade between the regions of the metropole and the conquered territories. Nevertheless, the great colonizers saw more to their immediate interests than to the future ones because the colonies, with the passing of time, would have adopted European-type productive models[6] and would have gained independence from the metropoles, and the advantages for Europe would thus come to an end. In fact, Genovesi saw that it was not unlikely that one day the colonies would surpass the mother country and, in turn, colonize Europe. "Everything in the world turns and all moves with the passage of time". There is no shortage of historical examples: "Could we Italians have ever thought at the time of Augustus that we could have become colonials of northern peoples?" (Genovesi [1765] 1824, p. 144).

At the end of the eighteenth century, less and less was said about sweet trade and the attention of economists shifted more to industrial activity and its relationship with violence too.

James Mill's thoughts on war, swinging from consensus to opposition in his early journalism at the time of the Napoleonic Wars, settled down in his mature period when he fell under the influence of Bentham's pacifist convictions. The pamphlet *Commerce defended*, published in 1808, stands between these two phases and defends free trade at the time of the continental blockade, opposing the statements made by William Spence and William Cobbett. In *Britain Independent of Commerce* (1807), Spence considers that the contribution of international trade to be insignificant in the formation of the country's wealth, while it benefits from war. For his part, William Cobbett, the founder and editor of the widely read weekly "The Political Register", saw foreign trade as the source of most wars.

Spence's reasoning on war is quite singular. He was worried that the tendency to save could excessively reduce demand and so hoped that the state would eliminate this excessive accumulation of money by taxation and public borrowing, going so far as to state that the higher the level of taxation, the higher the level of prosperity of producers, who would see all their production adsorbed. Since taxes are higher during wars, in those periods the population lives better (Spence 1807, p. 72). During the war, not only are producers at an advantage, but also the masses, because the young people who join the army are well fed and well dressed, while at the same time the call to arms reduces the availability of workers and causes wages to rise.

Even if showing that the very bases of Spence's reasoning are manifestly unfounded, as regards the role of trade both in the economy in general and in the evaluation of the effects of the war, Mill took it on himself to rebut the conclusions by the use of statistics that demonstrate the doubtless misery of the English poor during wars. Seeing that the measure of national prosperity is the standard of living of the working classes, he found it inconceivable that the devastating effects of war were not visible as it was the sole cause of economic stagnation and social misery. He considered that war derives simply from politically bad governments.

In the article *The Law of Nations*, Mill indicates the need to establish an international code of laws that defines the right of nations in both peacetime and wartime, so avoiding that the stronger nations always prevail. The code would also have to establish what condition a war must meet to be considered a just war. Mill defined the ways in which "the nations of the civilized world" (Mill 1820, p. 27) could agree to the stipulation of this code and the forms and procedures to be used to achieve the constitution of an international tribunal for arbitration *super partes* to prevent wars. This tribunal, which would hold public sittings, would not have coercive powers, because Mill was convinced, like Bentham, that the moral standing of such an institution would be so high in public opinion that it would force recalcitrant governments to bow to its deliberations. "The only sanction that can be applied to the Laws of the Nations is the popular sanction" (Mill 1820, p. 1).

The essay was influenced by a project for universal and perpetual peace written on the eve of the French Revolution by Bentham, but published only in 1843. In reality, however, Mill refers implicitly only to relations between "civilized" nations, while he accepts acts of war against "uncivilized" nations, not by chance justifying the English intervention in India where he thought that Great Britain could carry out works of civilization and economic improvement (Mill 1858).

Just what unleashes war? Certainly not international trade, as Cobbett holds, besides overestimating the effects it can produce. Mill recalls that trade allows buyers to purchase abroad at better prices than domestic ones and sell at a surplus profit, so increasing the wealth of the nation. This is profitable, but not the cause of war. The governments that state the opposite are not telling the truth. On the other hand, war certainly is harmful because it destroys the wealth of the nation. However, the prodigality of those states that waste wealth is equally harmful. The well-being of peoples develops where economic activity is free, private property is protected and peace guaranteed by the wisdom of peoples and sovereigns. In his mature years, Mill held that war is a brake on national growth, swallows up private savings, halts the country's progress and forms the source of economic stagnation and social misery. Colonization is among the causes that lead to war. This option is useless, even damaging, for the economy of the metropolis itself, because the conquest and maintenance of colonies turn out in the end to be more costly than the profit drawn from the conquered territory. Colonies cost and lead to monopolistic forms of trade of which privileged companies and a part of the class in government take advantage, to the detriment of the national community, which instead would have everything to gain from free trade with the colonies obtaining independence. A self-interested minority gains from the colonies, which leads the state to conquer them. Mill therefore speaks out forcefully against the privileges of these groups of overbearing people, asking for the colonies to be abandoned and free trade to be introduced, because the free play of economic forces encourages international solidarity. Good government of the economy does not require aggression and, so, its economic policy cannot be aimed at war.

John Stuart Mill did not write much on war and in any case nothing systematic. He did not develop the ideas of his father, even though he was decidedly against warmongering, even aside from considerations on the economic effects of war. He compared the devastating power of natural phenomena such as earthquakes, floods and hurricanes to the disasters produced by war when talking about the destruction of wealth and the capacity of countries to recover (Mill [1848] 1965, vol. 1, p. 74). Like his father, he thought international arbitrage was opportune in dealing with controversies between states. Unlike the mercantilists, he thought that the growth of international free trade was not just the source of economic well-being and a guarantee of peace, but also ensured progress in ideas, institutions and morality through the contact between different mentalities and habits (Mill [1848] 1965, vol. II, pp. 686 seq.).

John Stuart Mill's free trade views were deeply permeated by considerations of a political and moral kind. He believed that non-interference in the activities of foreign nations should be considered a "moral question" (Mill [1859] 1984, p. 118). Armed intervention was acceptable only when it served to promote the improvement of a country, meaning that a war may be undertaken only to liberate a nation oppressed by a foreign government, or to assist a backward nation in improving its economic conditions.[7]

## The Pacifism of the "Industrieux" and the "Industriels"

War belongs to the past was a conviction that many economists held from the second half of the eighteenth century onwards, being joined in the following century by positivist sociologists who equally held this view and emphasized it. The transition of power from the warlike nobility to the pacifist bourgeoisie engaged in productive activities, the expansion of commerce, free trade and, above all, the increase in labour productivity following the process of industrialization should have made warlike aggression superfluous, even anachronistic, things from the past which had been an important way to increase the wealth of states. War was relegated from first to last option as a solution to international conflicts, if it were not altogether impossible.

The determinist optimism of many classical economists and positivist sociologists was roughly interrupted by the return to colonization at the end of the nineteenth century, the outbreak of the First then the Second World Wars, followed by the permanent passage of new wars.

The technological innovations in agriculture in the seventeenth and eighteenth centuries, and more still, the industrial revolution starting in

Great Britain in the second half of the eighteenth century, focused the attention of economists on the increase of productive capacity first in the countryside and then, even more markedly, in factories. It seemed that for the first time in history it had become possible to overcome the dramatic limitation of the incapacity to push up production to meet the vital needs of a growing population. The optimism that characterized the century was due to the great hope that, thanks to the division of labour and the introduction of machinery, production could increase sufficiently to avoid famine or death from hunger and illness of the weaker part of the population following a single year of poor harvests. It was goods of primary importance that should grow in line with the growth of human needs and not humans that should have to adapt to available resources. Production and the low cost of goods would guarantee well-being not only to individuals, but also to the nation. Wealth would not have to be seized violently from others, but could be produced peacefully and increased domestically for civil consumption and exportation. War was to be avoided because maintaining large armies burdened the economy and fighting destroyed wealth, while economic cooperation among individual countries provided positive results for all concerned.

As we mentioned above, the mercantilists subordinated the economy to politics and the power of states, while laissez-faire economists wished to increase labour productivity and improve the population's living standards and so consequently the wealth of nations. The harmony of interests would be achieved automatically, not only between the population and the state, but also at an international level, so making peace possible. The physiocrats saw this happening on the basis of the laws of nature that sovereigns were called on to observe without interference, while for the laissez-faire economists it was the laws of the economy that led to harmony. In fact, for some of them peace was the true objective of the economy and the task of economists was that of publicizing laissez-faire theory so that its concrete application would lead to peace.

Physiocrats and laissez-faire economists agreed in seeing that domestically realized wealth obtained through productive activity was much more advantageous than that got from wars against others that cause devastation and poverty for the vanquished and were a cost to victorious nations too (Quesnay [1759] 1888, p. 242). Wars redistribute wealth, destroying a part of it, but peaceful production increases overall wealth to the benefit of all. Not only warlike conflicts, but also trade wars and attempts at usurpation were to be considered as negative and harmful to general well-being.

Some, like Dupont de Nemours, started out with the political objective of human happiness. This can be realized just by having citizens free to act as they wish, with guarantees of personal safety and private property of their possessions, conditions that do not obtain in times of war. Consequently, following moral or religion, the right to unleash war could belong to no nation (Dupont de Nemours 1792, p. 2). He maintained faith to this idea when, as President of the National Constituent Assembly in 1790, he presented the draft of a decree with the first article stating: "The French nation will not permit an offensive war to obtain the territory of others, nor to limit the rights and liberties of any nation".

As for the British classical economists, neither Smith nor Ricardo paid special attention to the theme of war as a phenomenon, as it was implicit that industrial development and free trade would lead to the harmonization of interests and international pacification. Both Smith and Ricardo saw that economic freedom guaranteed nations the possibility to gain wealth peacefully through production specialization. Ricardo thought that wars were waged to favour the interests of private groups, while citizens are burdened with the payment of useless costs to pay for them. The interests of nations are not antagonistic and so should not lead to conflict, and capitalism is pacific by nature as it is transnational.

Nevertheless, there was still the question of defence, which Smith dealt with in the fifth book of the *Wealth of Nations* when considering the management and control of public possessions which are not subject to the laws of the market, that is what we today call the public economy. Smith's list of activities to be financed by public money is headed by defence, because the sovereign's main duty is to defend society from violence and aggression by "other independent societies", and this requires the maintenance of a military force. War is a threat to citizens, the army and the market economy. Defence, whose task is to guarantee international security, carries out a social function from which all society gains. The cost of maintaining it is therefore not defined on the basis of the laws of the market and so it is normal that all take part, if possible, in proportion to the means they have to hand.

The expense for defence, both that of peacetime preparation and that of use in wartime, turns out to be very different according to different types of society. In "opulent and civilized nations", the cost of armaments is much higher than in "poor and barbarous" ones, while the arms of civilized societies are much more efficient (Smith [1776] 2007, p. 472), because the art of war, with the passing of time and technological progress, has become a science and has become increasingly complex. The economic laws, which regulate production, exchange and the redistribution of wealth, do not appear to apply to military activity. The choice between a militia and a standing army is not made on the basis of lower cost.

Leaving aside the needs for defence, Smith did not think that a country had to accumulate gold or silver to be able to wage war abroad and to maintain fleets and armies in distant countries, because "fleets and armies are maintained, not with gold and silver, but with consumable goods" (Smith [1776] 2007, p. 238).

Smith does not clarify the matter of potential aggressors in his own world, but we have to guess that he thought of nations that were not very civilized, poor and in any case uncompetitive economically, which still thought of gaining wealth by obtaining war loot.

In the case of financing a war, it was better to lay on direct taxes rather than to take out loans that put off payment. With an immediate increase in taxation, the population more directly feels the burden of the war and soon wearies of it, forcing the government not to protract it beyond the minimum time required. Furthermore, "The foresight of the heavy and unavoidable burdens of war would hinder the people from wantonly calling for it when there was no real and solid interest to fight for" (Smith [1776] 2007, p. 605). However, Smith does not tell us what type of real and solid interest can justify a war.

Ricardo advised governments not to engage in wars at the expense of the public treasury and focused on the examination of the effects of the Napoleonic Wars on the British economy. In his *Principles of Political Economy*, he dedicated a chapter to the economic consequences of sudden changes that can be caused by the outbreak of war after a long period of peace or a peace that follows a long period of war. Money lies unused and so causes unemployment during the time required to organize the reconversion and redirecting of capital from sectors of production in peacetime to those of wartime, or vice versa. War breaks off normal trade relations and can cause problems for the food supply because it hinders, among others, the import of grain. The agriculture of a nation at war, in this case Britain, is consequently forced to produce more, albeit at a high cost. When the conflict ends and trade restarts, domestic cultivators find difficulty in cutting back production and special protective tariffs have to be introduced to remain in force only as long as strictly necessary for producers to switch their capital from wheat production to other sectors. If a war broke out in continental Europe, on the other hand, Ricardo held that Great Britain would still be able to purchase the grain it required from a neutral country. Fear of war should therefore not give rise to a preventive agricultural protectionism. The supply question did not arise for the industrial area, given the British superiority in the sector over the rest of Europe.

In the third edition of the *Principles* published in 1821, Ricardo considered the influence of war on wages and employment too. He held above all that the higher taxes on wealthy people during the war to pay for the army provided means of subsistence for a higher number of people than that who could benefit from the money being left to the contributors. War is recognized for the merit of an increased demand for labour, at least.

As regards financing, Ricardo too considered that the cost of war should be covered by taxation and the best means to guarantee peace was to force ministers to ask personally the taxpayer to pay new taxes to cover the cost of war (Ricardo 1821). According to Ricardo, wars are often unleashed for private interest rather than to the advantage of the country, while keeping the peace in Europe was "a great object" (Ricardo [1811–1823] 1899, p. 213) because the real interests of the different nations were not at all antagonistic.

Malthus saw war as a factor limiting the growth in the population of a country in his reflections on the problem of overpopulation compared with available resources. Nevertheless, war was not desirable and preventing excessive population growth through persuasion to reduce the birth rate could serve instead, avoiding wars to brutally reduce it.

Malthus witnessed the expansion of the British economy during the Napoleonic Wars and later the crisis after peace was declared in 1815. He sought an answer as to why this situation arose, reaching the conclusion that at the end of the conflict, the reduction in public spending and war taxation had created a corresponding increase in private saving and a reduction of overall demand, leading to depression. Malthus still recognized that the economic advantages that could be attributed to the war in fact derived from an artificially bloated demand by the state and that these advantages were passing and obtained at the price of major and lasting suffering by the population.

As is known, Malthus included soldiers among the unproductive classes. His ideal society was peaceful and resolved the problem of overpopulation by reducing the birth rate, precisely to avoid war intervening to reestablish the balance between resources and population. The principal cause of war was, in fact, due to the scarcity of food or land to grow it. In other words, the poverty of the population causes wars to break out. Malthus did not take into consideration the hypothesis of such vast economic and technological progress as to resolve permanently the problem of food scarcity and poverty in general, thinking rather that the population should continue, as in the past, to adapt to resources and not vice versa. If the population grew less, there would no longer be famine and wars would not break out.

Jean-Baptiste Say in his principal works (*Traité d'économie politique* and *Cours complet d'économie politique pratique*) displayed a total opposition to warmongering and described the devastation produced by war, admitting only the need for defence, because peace is essential for the economic development of countries. Technical progress had made war much more expensive than in the past: the cost of armaments during the fighting, but also in peacetime, eats up savings and cuts back national capital too. In any case, not even the most powerful arms served as a deterrent and could not save a country from war either. Defence did not require huge expensive armies as no army had ever covered a country against invasion. On the contrary, it was precisely major preparation for war that led to the outbreak of armed hostilities.

War is all the more devastating when the theatre of operations is full of industrial plants and its real cost is always higher than the cost for fighting it because the losses for the war must include other factors. One is the cost of keeping and instructing the young people who lose their lives in the fighting before they can return to the community, through work, the value corresponding to their training. Another cost is that represented by the value of what the war effort detracted from producing works of peace. This shows that Say came to the concept of "opportunity costs" a century early.[8]

Hope springs eternal. Seeing that war becomes increasingly expensive, only the rich can afford to wage it. Usually, these are the more civilized countries, which, as such, would not wish to unleash wars. Equally, usually civilized countries are democratic too, so they must have the approval of their citizens before declaring war and they will obtain this increasingly less often as the masses increasingly see where their real interest lies. A contribution to distancing war from history is also offered by progress in political economy, the science whose objective is just the realization of peace (Say 1803).

Preponderance is not economically useful even for those who exercise it. In the past, governments and peoples made war to conquer a city or a province, to monopolize a branch of commerce and to win colonies and then hold on to them. They would have gained a greater benefit from friendly relations as it is not in the interest of those who produce wealth, farmers and industrialists alike, to go and devastate foreign lands, seeing the risk they are running of being attacked from outside due exclusively to the errors or the "passion" of their governments. The latter work hard to make the wars they intend to wage in the interest of a few "inevitable" (Say [1803] 2008). Those who push for war are military people or politicians who hope to obtain a personal advantage, interested minorities in either case. The majority of citizens, on the other hand, accept war of aggression following a misunderstood economic interest, on which political economy must cast light.

Say's epoch was characterized by the emergence of bourgeois values and economic interests, with the start, in France too, of the process of industrialization and the consequent growth of the *industrieux*, a social group composed of manufacturers, tradespeople and workers, but also including farmers and other "wise people", all intending to produce wealth, doing so all the better in peacetime. Their peaceful attitude was the opposite of the warlike and predatory spirit that dominated the military, who produced nothing, but instead appropriated the product of others. Say saw the soldier, differently from the unproductive worker in Smith, as a destroyer of wealth.

On winning political power, the *industrieux* would eliminate protective economic systems and pursue a liberal, laissez-faire policy inspired by the principles of international solidarity. These principles, as defined by economic science and which, for Say, form the very foundation of morals, of both individuals and nations, were finally put into practice, thereby determining the end to war. Warmongering constituted the natural state of mankind only when it was ignorant of the precepts of economics and consequently was unable to evaluate correctly their own interests.

Bastiat's analysis of war started from the examination of the two ways mankind can acquire the means of survival: production and seizure. The second is achieved within the country by cunning and violence, and outside the country with war, conquest and the colonies. The will to acquire the labour of others is the origin of armed conflict between countries, but only monopolists gain an advantage from wars and do so to the detriment of the masses. If it is possible to export goods at slightly below the price of foreign goods, there is no need for guns and ships to enrich the country. Governments, instead, follow a totally irrational economic policy by taxing raw materials, means of production and even consumer goods, making prices rise, then raise other taxes to arm the navy and send it to find trade outlets abroad. Everything would be so much easier, effective, peaceful and less costly, if there was a free regime of production and trade.

Bastiat was very active on the pacifist front. In 1849, when he was a people's representative at the National Assembly, he published a tract called *Peace and Freedom* calling for international pacification, disarmament and the simultaneous reduction of taxes to create general well-being. In the same year, he went to the Second Congress of the Friends of Peace held in Paris along with Michel Chevalier, Charles Dunoyer and Gustave de Molinari. He launched an appeal from the platform, proposing that countries should live in peace, because their interests were harmonic and the antagonism that led them to fight was only apparent.

Frédéric Passy, Bastiat's follower, also started from a supposition diametrically opposed to the mercantilist one when he thought that economic evolution determined politics and not the other way round. Passy saw that the mercantilist and protectionist concepts were the principal source of wars and held that, sooner or later, they would be banned as crimes against humanity. In the meanwhile, free and frequent trade relations between countries led to maintaining peace. This is where we can find the greatness, the truth, the nobility and "almost the holiness" of the doctrine of free trade, which, thanks to the prosaic but effective pressure of interests, tends to make justice and harmony in the world prevail (Passy 1862, p. 582). Economic well-being can only exist in times of peace. War, for Passy, was not only a crime, but also an absurdity.

Passy was a convinced and active pacifist just as much as Bastiat and in 1867 founded the "Ligue International et Permanente de la Paix"; then, in 1901, he won the Nobel Peace Prize.[9]

However, in France some laissez-faire economists did not fully agree with the optimism of Say, Bastiat and Passy concerning a rapid achievement of universal pacification. The analyses of Pellegrino Rossi, Michel Chevalier, Henri Baudrillart and Gustave de Molinari took political factors into greater account and demonstrated less willingness to allow economic science to fly in the face of the reality of facts with optimistic speculation (Rossi 1850, vol. 2, p. 208). These economists were also convinced that the economy alone could not modify the path followed by the world by offering solutions to the problem of wars. Pellegrino Rossi, for example, held that the internationalization of politics and the economy was an objective that could not be realized concretely. He thought instead that national states would continue to exist as the indispensable means for the human species to achieve material prosperity and moral improvement. However, if states continued to exist with their political reasons, imagining the world as one vast market, the premise for peace, among other things, it could seem to be an abstraction, or at least a hope, that would be difficult to realize.

Michel Chevalier, who negotiated the Anglo-French Free Trade Agreement in 1860 with Cobden, saw a pacifying tendency in industrial civilization, but did not believe that a free market alone would lead to peace, preferring to think that free trade presupposed the existence of a situation of international peace. For him, peace was the result of scientific and industrial progress, realized within a representative political system. Nevertheless, he did not think that permanent peace was possible as the aggressive instinct formed part of human nature. Consequently, economic analysis had to take into account the possibility of wars breaking out between states, which therefore could not demobilize armies for defence.

Henri Baudrillart too (1860) held that political economy, with its laissez-faire precepts, was not able alone to obtain the elimination of war. However, he thought that it could effectively fight off the prejudices that justified war by denouncing the economic damage that it caused for the winning countries too. Even if he agreed with the idea that the development of production drew countries closer to peace and that industrialization created an international community of interests, nevertheless, the perspective of permanent peace appeared to be still far off.

Gustave de Molinari, as Alan Milward recalls (1977, p. 2), was "One of the few economists who tried to integrate the existence of war into classical economic theory", considered the increase in the cost of waging war over the centuries and wondered if the winners could still gain an economic advantage from war.

Molinari shared the idea of the classical economists, who thought that the wars in the past were dictated by the lack of economic resources that could only be covered by violent seizure with sacking. Even if violence is necessary for the survival of animal species and was the case for primitive men, it is no longer so, for in contemporary developed society where, thanks to the increase in labour productivity, men can obtain what they need without attacking other men. Modern wars cannot offer good profits, as in fact they turn out to cause damage not only for the losers, but also for the winners too, because the costs of the war are higher than the profits which may be made from the victory, but as Molinari wonders, if this is the case, why do wars still break out? Who gains from militarism in contemporary society? Molinari, like many who had previously asked the same question, also finds in the men in government and the military career elite those social groups who can gain a personal advantage from a victorious war. Only this very small minority, which provokes wars by taking advantage of militarism, the most costly of anachronisms, stands to gain. The other classes and the common people have to put up with destruction and financial costs.

According to Molinari, in the nineteenth century and in all states, regardless of their constitutional set up, power was still exercised by a class interested in the persistence of war and the enormous apparatus for destruction that it required. This was because the masses, who were interested in peace, were taken up in daily productive activities and still did not have the necessary influence to force governments to renounce war. Nevertheless, the interests of the classes in favour of peace were progressively becoming more important, and in time, these classes would come to exercise an influence in proportion to their economic importance and would be able to convince public opinion to oppose warmongering.

Molinari suggested the constitution of an international organization to be called the League of Neutral Countries to speed up this process and imagined that it could take form with an agreement between Great Britain and the small states of continental Europe, which, just because they were small, could run the risk of seeing their autonomy under threat. The Netherlands, Switzerland and Denmark could have an interest in forming a coalition. The League of Neutral Countries could act as a peacemaking buffer force. Molinari drew up this project in 1870, the year of the Franco-Prussian War, but also on the eve of the long depression at the end of the nineteenth century and the massive return to colonialism.

Molinari's theory of the need for peace drew on elements from Say, Saint-Simon and Comte and saw war as a necessary factor, as well as useful and moral, in the past (considering wars as a means of opposing barbarians), but superfluous, harmful and immoral in the present. Molinari's peace was obviously linked with free trade.

While French economists thought of establishing pacifist associations alongside the popularization of laissez-faire ideas, the English preferred the use of law. As we have seen, Bentham and James Mill were convinced that the balance of power between countries should be substituted by rules dictated by law and enforced by specially appointed international organizations and tribunals.

Pareto held Bastiat and Molinari in high esteem, but wrote little about war. In his youth, he held conferences, attended seminars and published pacifist articles accusing protectionism of being intrinsically warlike during his period of free trader in the late nineteenth century. In his *Cours d'économie politique*, he recognized that men could obtain all that was necessary through production and transformation of economic goods or, alternatively, through the appropriation of goods produced by others. Harking back to Molinari, he placed war among useless institutions that have outlived their usefulness and, instead of loosening their grip, have increasingly held on tight, as they became increasingly useless (Pareto 1896–1897, vol. II, pp. 44–45).

More generally, he established the economic analysis regarding the conditions which define the price of equilibrium which excludes the factor of war. Pareto and his mathematical school viewed conflicts as a dynamic element outside the economy (cfr. Wright [1942] 1960, p. 1369). It is just where political economy gives way to the application of mathematical methods for the formulation of theories and models, in particular models of general economic equilibrium, that war is removed from the field of investigation, at least until the emergence of game theory.

Though Pareto, like Walras, did not take war into consideration in the mathematical methods he used for the economy, he mentioned it in his sociological research. His theory of elites considered the irrational motivations of human behaviour and placed war among the "residues" and the "derivatives" as symbolic manifestations that act in social behaviour in which elites intervene to their own profit. War derives from the upper hand held by the "residues" and is carried out by the "lions", that is the elite which uses force and believes in the persistence of "aggregates", that is, believes in its ideals and is ready to use force to realize them. Gradually, the maintenance of the residues and the aggregates enters into a crisis and is superseded by the analytical and rational mentality of the "wolves" (the elite that uses cunning). However, with the elimination of the aggregates, the road is clear for a new group of "lions" which brings forward new ideas and leads to new wars.

## The Contribution of Sociologists

The theses on pacifism in industry formulated starting in the second half of the eighteenth century with the beginnings of the industrial revolution in Great Britain gained its greatest audience at the beginning of the nineteenth century through the work carried out by newly born sociology.

It was the sociologists and the positivists who strongly supported the idea that the new industrial society, that was beginning to dominate in Europe in the first half of the nineteenth century, was intrinsically pacific. Just as for the laissez-faire economists, so too for the positivist sociologists, economic development would eliminate war as military spirit was incompatible with industrial society. There was an antimony between war and work. The progress of science and technology would lead to a constant growth in productive capacity and the market economy would realize general well-being and eliminate social tensions and those between states. The result would be a new era of peace and prosperity.

Saint-Simon and Comte held that industrialization had radically transformed society and, thanks to the reduction of the power of military institutions and the elimination of rivalry between states, this would lead to the dominance of a peaceful capitalism.

Saint-Simon started in the mid-nineteenth century with the construction of a philosophy of history based on the counter-position of "the industrial spirit" to the "military, or feudal, spirit" and the inevitable victory of the former. The instinct of domination in its primitive and military form, implying the violent conquest of territory and the enslavement of the vanquished, was necessarily succeeded by the industrial spirit, thanks to which men associate to dominate together the forces of nature, while seeking not to dominate each other. "Industrialists" meant everyone who contributed in some way to the production process, including the students of "positive sciences" and the artists and lawyers too who defended the industrialists, as well as some clergymen. In the other camp, there were the parasites who lived off others' work: the nobility, many clergymen, owners of real estate and, obviously, the military. The new industrial class had to take on the organization of society by itself, eliminating the bureaucratic government structures. The army, based on blind obedience, was to disappear with the arrival of the industrial system and, with it, wars.[10]

Wealth did not increase through violence in contemporary society. Saint-Simon (1869), like Say, thought that war was harmful to all, for men, for industry and for trade. He held that the growth of industrial spirit and technological innovation were the guarantees for future peace, unlike monopolies and colonies. Saint-Simon shared with Say the idea that warmongering and well-being cannot coexist and wars, whatever their aim, cause damage to humanity. The thought that national wealth can be increased by war was a prejudice feeding on ignorance. War hinders production and blocks trade. The guarantee for future peace lay in the expansion of the "industrial spirit" and it is only because this industrial spirit did not yet prevail that national hatred continued to exist and the consciousness of the community of interest was still to be gained. Saint-Simon published a pamphlet with his follower Augustin Thierry in 1814 in which they upheld, rather confusedly, that in order to eliminate wars it was vital to eliminate national states to form a super-national political body.

The necessary historical transition from military society to industrial society was an idea shared and theoretically formulated by August Comte, the father of sociology, who in 1842 greeted the arrival of a new era in this way: "finally we have arrived at the epoch in which effective and permanent war must disappear altogether from the best part of humanity" (Comte [1830–1842] 1967, vol. II, less. LVII).[11] Humanity had to praise the laissez-faire economists with "eternal recognition for their happy efforts intent on dissipating the sad and immoral prejudice that, both between individuals and between peoples, represented the bettering of the material condition of one as only possibly arising from the corresponding worsening for the other" (Comte [1830–1842] 1967, vol. I, p. 189).

The civilization of work would substitute the civilization of arms. Wars in the past were due to the refusal of work by the social classes in power, which instead concentrated on war operations: "in antiquity work was for slaves, while freemen were available for war. The primitive disdain for work made the use of war inevitable" (Aron 1957, p. 6). The relationship would have to invert slowly in contemporary society with the predominant activity no longer being war, but free work, which would replace slavery. Consequently, disdain would be for war.

"Trade" wars, those with an economic goal, and colonial and imperialist wars too would have lost their role thanks to the improvement in the productivity of labour. Phenomena not in line with the radical change in society, like wars, would inevitably vanish with time. Since the ruling class defines the hierarchy of social values, industrial society, by taking over from military society, guarantees the transition to peaceful prosperity.

Aron noted that, as an irony of history, Comte's prophecy had been half realized; engineers had taken over from the nobility in handling economic power, but the best engineers worked for the armaments industry; industry had expanded massively, but it had not become peaceful and it was war that industrialized itself, not industry that pacified the world.

Comte's reading of the process of industrialization historically falls into three phases, the first of spontaneous and autonomous development, the second of state intervention in favour of industrial growth, seen as the means to gain military supremacy, and then finally the third in which industrialization had become the aim in itself in the policy of European states. We should recall that for Comte the term industry meant every productive activity requiring the employment of free labour, be it agricultural, manufacturing or industrial in the real sense of the word, or the provision of services.

Spencer too made a distinction between military society and industrial society, and in his case too, the term industrial has the wider meaning including commercial and banking activities (and, in general, all non-warlike activities), which also contribute to individual well-being. Nevertheless, for Spencer, the evolution from the warrior type of society to the industrial one is not inevitable and may not be linear. He saw that rather it was historically necessary to undergo an evolution from homogeneous societies to heterogeneous societies (a thesis later adopted by Durkheim). The outbreak of war was still a possible event and could lead to drawbacks in the process of evolution with the return to forms of military organization of the state and society. On the other hand, an industrial typology can be found even in only partly advanced societies, which, nevertheless, know a type of productive cooperation characteristic of industrial societies. There can also be an interchange between the two models, as was happening in Great Britain, according to Spencer. As he saw it, in order for industrial society to emerge, wars had to disappear, while for Comte the opposite would take place: the emergence of industrial society would lead to the elimination of war, which would lose its function and become useless. For war to disappear, Comte said, firstly philosophers should denounce the evil it brings.

The First World War cancelled the optimism of the sociologists. It was yet another war for hegemony, with Germany seeking to conquer territory in Europe and the colonies. Spencer could read it as a major phase of regression, and others, like Burckhardt, could see it as the beginning of a new "military industrial" phase in which just the condition of advanced industrialization led to the war. Andrea Graziosi wrote on this: "because of the need for power, imposed by the development of industry and then above all by the Spencerian regression caused by the war, the twentieth century has offered us, for example, more than one type of the 'military industrial' state clearly foreseen by Burckhardt" (Graziosi 2001, pp. 37–38).

The analysis and judgement of some leading sociologists regarding war changed radically after the First World War. Back in 1915, following the invasion of Belgium, Emile Durkheim in *Germany above all* explained German nationalism through the analysis of the writings of Heinrich von Treitschke, a nineteenth historian who was a friend of Bismark, a pro-imperialist and Pangermanist whose thought can be considered as being representative of the common German feeling before and during the war.

According to Treitschke, the state stands above international law, morals and civil society and has the precise duty to be strong. Its strength and power are expressed firstly through the army, the cornerstone of society and the "incarnation" of the state itself. Permanent peace is to be avoided because it is a synonym for weakness and egoism, while war is a generous and useful act, which encourages a feeling of belonging in the population. Without war, the state cannot even be conceived.

Durkheim considered the will to power expressed by Treitschke as "a diseased hypertrophy of the will, a kind of mania to want" that had involved an entire nation in a certain historical moment (Durkheim 1915, p. 43).

Max Weber, in his two volumes *Economy and Society* published in 1922, held, among other things, that the enemy was indispensable for collective cohesion and the very affirmation of the state and recalled that all political formations made and still make use of force externally.[12] In particular, the great powers, just because they are great, are often "expansive", that is "they are associations aiming at expanding the territories of their respective political communities by the use of violent means or the threat of force" (Weber [1922] 1978, vol. 2, p. 912).

Then, there are obviously private economic interests that aim at war, above all armaments manufacturers and those who finance them. Free trade, which is peaceful and non-monopolistic, can certainly affirm itself, but only as long as the private organization of covering requirements is profitable. When this ceases to be the case, there is the "universal revival of imperialist capitalism" which, according to Weber, has constituted over time the normal form of capitalism's influence over politics, pressing for conquest (Weber [1922] 1978, vol. 2, p. 919). When Weber was writing, everything led to foresee that this imperialist capitalism was going through a long period of affirmation.

Raymond Aron was the main writer to reflect on war after the Second World War and linked it above all to political factors, making him one of the founders of the "realist" current. Aron looked above all at the contemporary condition he found himself in, that of the Cold War, in which the tension between the two main powers had led to a situation in which war was improbable and peace impossible. In his *Paix et guerre entre les nations* (1962), Aron made a general excursus on the interpretations the economic, social, historical, political, moral and strategic literature on war, seeking to establish among them precepts for action that turned out to be rational in theory and reasonable in practice. He concluded that there were no general rules capable of orientating international relations towards pacific solutions, there were no infallible solutions and there was no univocal and non-dogmatic scientific response. Instead, the situations had to be examined as they emerged historically, case by case. Aron sought a third way between the dream of a universal republic and the permanent struggle between states that would be dictated by ethics of responsibility and would be made possible by the balance of power, the way which would allow national interest to be preserved while war was prevented or, at least, limited in frequency and violence (cfr. Le Bras-Chopard 1994, p. 145). The objective Aron considered to be viable and possible for every individual was that of: "not to run away from a belligerent history, not to betray the ideal; to think and to act with the firm intention that the absence of wars will be prolonged until the day when peace has become possible – supposing it ever will" (Aron 2009, p. 787).

The problem of the relations between states, war and capitalism was taken up again in the 1980s by Michael Mann in research in political sociology in his papers collected in a single volume in 1988. Here, Mann stressed how previous sociological analyses had erroneously separated the study of relations between capitalism and war from that between states and war, which are closely correlated. This will be dealt with again below.

## Notes

1. "Ce fut la raison principale que meut Scipion le Jeune d'empescher tant qu'il peut, que la ville de Cartage fust rasee: prevoyant sagement que si le people Romain guerrier et belliqueux, n'avait plus d'ennemis, il estoit force qu'il se fist guerre a soy mesme."
2. Hermann Moritz von Sachsen (1696–1750), better known as the Maréchal de Saxe, was Marshal of France and theorist of the art of war. He wrote his *Méditations* in 1732, which were published posthumously.
3. Some mercantilists, however, took account of this need and Thomas Mun, a leading English merchant in the fifteenth century, held that a sovereign had to develop foreign trade in order to accumulate wealth, but must equally "enrich his own subjects, favouring his own interest in that way. In fact, a prince's power is evaluated more on the basis of the wealth and the dedication of his subjects than on the basis of the quantity of money he holds in his own treasury" (Mun [1664] 1928, p. 69).
4. István Hont published a large volume in 2005, republishing his previous studies, with the same title *Jealousy of trade*, retracing Hume's "jealousy of trade" back to the evolution of the "Jealousy of Kings and Persons

of Sovereign Authority" or, in short, "jealousy of state" mentioned by Hobbes in the XIII chapter of *Leviathan*. Hont starts from this change of the object of "jealousy" in his study of the genesis of the change taking place in modern politics (Hont 2005, pp. 1–2).

5. Gallagher and Robinson (1953) studied the formal and informal imperialism of Great Britain in the first and second phases of the Victorian era.
6. This statement foresees the reflections of Rosa Luxemburg on the transformation colonization made on the mode of production in the colonized countries.
7. See Philippe Gillig (2018).
8. Say estimated that Napoleon's wars cost humanity about ten billion francs.
9. Perhaps, it is significant that a century later, Aumann, a trained mathematician, won the Nobel Prize for Economics for work starting from the presupposition that war is a rational activity.
10. The writings of his followers differed, in particular those of Prosper Enfantin, who believed that industrial peace derived from military organization.
11. "Permanent war" was indicated by George W. Bush as a new condition of contemporary society.
12. Apparently, only the Inuit, alone in the world, do not have a concept of war and do not even have word to describe it.

## References

Anonymous [William Petyt?]. 1680. *Britannia Languens, or a Discourse of Trade*, in Classical Writings on Economics as Selected by J. R. McCulloch. Vol. I. London 1995, pp. 291–504.

Argenson, René-Louis de Voyer de Paulmy, Marquis d'. 1764. *Considérations sur le gouvernement ancient et present de la France*. Verdun: BnF Gallica.

Aron, Raymond. 1957. "War and Industrial Society." *Auguste Comte Memorial Trust Lecture 3*. London: Oxford University Press.

———. 1962. *Paix et guerre entre les nations*. Paris: Calmann-Levy.

———. 2009. *Peace & War. A Theory of International Relations*. New Brunswick and London: Transaction Publishers.

Bacon, Francis. [1625] 1944. *The Essayes or Counsels, Civill and Morall.* Oxford: Clarendon Press.

Bastiat, Frédéric. 1849. *Paix et liberté ou le budget républicain*. Paris: Librairie de Guillaumin et Cie.

Baudrillart, Henri. 1860. *Philosophie de l'économie politique. Des rapports de l'économie politique et de la morale*. Paris.

Bentham, Jeremy. [1838–1843] 1873. *The Works of Jeremy Bentham*. 11 voll. Edinburgh: John Bowring.

Bodin, Jean. [1568] 1932. *Réponse à M. de Malestroit*. Paris: Colin.

———. [1576] 1986. *Les six livres de la république*. Paris: Fayard.

Boisguillebert, Pierre Le Pesant, Sieur de. [1707] 1966. *Traité de la nature,culture, commerce et intérêt de l'argent* in *Pierre de Boisguillebert ou la naissance de l'économie politique*. Paris: Institut National d'Etudes Démographiques.

Botero, Giovanni. [1589] 1956. *The Reason of State*. London: Routledge & Kegan Paul Ltd.

Cobbett, William. [1807] 1998. *The Collected Social and Political Writing of William Cobbett*. 17 voll. London: Routledge.

Cobden, Richard. 1870. *Speeches on Questions of Public Policy*, vol. 2. London: Macmillan.

———. 1878. *Political Writings*. London: William Ridgeway.

Colbert, Jean-Baptiste. 1861–1882. *Lettres, instructions et memoires de Colbert*. Paris: Imprimerie Nationale.

Comte, Auguste. 1830–1842. *Cours de Philosophie Positive*. Various Publishers. Also in: BnF Gallica. Edition consulted: *Corso di filosofia positiva*. 1967. Torino: Utet.

Dupont De Nemours, Pierre-Samuel. 1769. *Du Commerce et de la Compagnie des Indes*. Amsterdam. BnF Gallica.

———. 1792. *De l'Amour de la Constitution et de celui de la Liberté*. Paris: Imprimerie de Dupont.

Durkheim, Émile. 1915. *La Germania al di sopra di tutto. Il pensiero tedesco e la guerra*. Paris: Librairie Armand Colin.

Gallagher, John, and Roland Robinson. 1953. The Imperialism of Free Trade. *The Economic History Review* 6 (1): 1–15.

Genovesi, Antonio. [1765] 1824. *Lezioni di Commercio o sia di Economia Civile*. Milano: Società tipografica de' Classici Italiani.

Gilling, Philippe. 2018. "Economic Non-intervention and Military Non-intervention in John Stuart Mill's Thought." In *War in the History of Economic Thought*, edited by Yukihiro Ikeda and Annalisa Rosselli. London: Routledge.

Graziosi, Andrea. 2001. *Guerra e Rivoluzione in Europa 1905–1956*. Bologna: Il Mulino.

Hont, István. 2005. *Jealousy of Trade*. Cambridge, MA: Harvard University Press.

Huet, Pierre-Daniel. 1712. *Le grand trésor historique et politique du florissant commerce des Hollandois», dans Tous les états et empires du monde*. Paris: F. Fournier.

Hume, David. [1752] 2000. *Political Essays*. Milano: Cofide.

Infantino, Lorenzo. 1980. *Sociologia dell'Imperialismo: Interpretazioni Socialiste*. Milano: Franco Angeli.

Le Bras-Chopard, Armelle. 1994. *Théories et ideologies*. Paris: Montchrestien.

Malthus, Thomas Robert. [1798] 1826. *An Essay on the Principle of Population*. London: John Murray.

Mann, Michael. 1988. *States, War and Capitalism*. Oxford: Basil Blackwell.

McCulloch, John Ramsay. 1856. *Selected Collection of Early English Tracts on Commerce*. London: Political Economy Club.

———. 1872. "Introductory Discourse. Rise and Progress of Political Economy Down to the Publication of the *Wealth of Nations*." In *An Inquiry into the Nature and Causes of the Wealth of Nations by Adam Smith LL.D*, XV–LVI. Edinburgh: Adam and Charles Black.

———. 1995. *Classical Writings on Economics as Selected by J.R. McCulloch*. 6 voll. London: William Pickering.

Melon, Jean-François. [1734]. *Essai Politique sur le Commerce*. In *Collection des principaux économists*. BnF Gallica, translated by *David Bindon 1738. A Political Essay Upon Commerce*. Dublin: Philip Cramton.

Mercier de La Rivière, Pierre Paul. [1767] 1910. *L'Ordre naturel et essentiel des sociétés politiques*. Paris: Librairie Paul Geuthner.

Mill, James. 1808. *Commerce Defended. An Answer to the Arguments by Which Mr. Spence, Mr. Corbett and Others, Have Attempted to Prove that Commerce Is Not a Source of National Wealth*. London: C. and R. Baldwin.

———. 1820. *Law of Nations*. Supplement to the *Encyclopaedia Britannica*. Fifth edition.

———. 1858. *The History of British India*. 10 vols. London: Piper, Stephenson and Spence. Reprint by Routledge/Thoemmes Press in London, 1997.

Mill, John Stuart. [1848 ] 1965. "Principles of Political Economy." In *Collected Works of John Stuart Mill*, edited by J. M. Robson, vol. II. London: Routledge & Kegan Paul.

———. [1859] 1984. "Essays on Equality, Law and Education." In *Collected Works of John Stuart Mill*, edited by J. M. Robson, vol. XXI. London: Routledge & Kegan Paul.

Milward, Alan Steele. 1977. *War, Economy and Society, 1939–1945*. Berkeley: University of California Press.

Molinari, Gustave de. 1849. *Les soirées de la rue Sain-Lazare*. Paris: Guillaumin et Cie.

Montchrestien, Antoine de. [1615] 1899. *Traité de l'économie politique*. Paris: P. E. Plon, Nourrit et Cie.

Montesquieu, Charles-Louis de Secondat de la Brède et de. [1748] 1950. *L'esprit des lois*. Paris: Nagel.

Mun, Thomas. [1664] 1928. *England's Treasure by Forraign Trade*. Oxford: Blackwell.

North, Dudley. 1856. "Discourses Upon Trade: Principally Directed to the Case of the Interest, Coynage, Clipping, Increase of Money." In *Selected Collection of Early English Tracts on Commerce*, edited by J. R. McCulloch.

Palmieri, Giuseppe. [1788] 1805. "Riflessioni sulla Pubblica Felicità relativamente al Regno di Napoli." In *Scrittori Italiani di Economia Politica, parte moderna*. Milano: Stamperia e Fonderia G.G. Destefanis.

Pareto, Vilfredo. 1896–1897. *Cours d'Économie Politique*. Lausanne: Rouge.

Passy, Frédéric. 1862. *Leçons d'Économie Politique faites à Montpellier*. Paris: Librairie Guillaumin et Cie.

Quesnay, François. [1759] 1888. *Œuvres Économiques et Philosophiques*. Francfort et Paris: Oncker.

Ricardo, David. [1811–1823] 1899. *Letters of David Ricardo to Hutches Thrower and Others 1811–1823*. Oxford: Clarendon Press.

———. [1821] 1971. *On the Principles of Political Economy, and Taxation*. Harmondsworth: Penguin.

Rossi, Pellegrino. [1840–1851]. *Cours d'économie politique*. Bruxelles: Société Thypographique Belge.

Saint Simon Claude-Henry de Rouvroy comte de. 1869. *Œuvres*. II: *L'industrie*, III: *Du sistème industriel*. Paris: Dentu. Anastatic edition 1966. Paris: Edition Anthropos.

Saint- Simon Claude-Henry de Rouvroy comte de, Thierry Augustin. 1814. *De la réorganisation de la société européenne ou de la nécessité et des moyennes de rassembler les peuples de l'Europe en un seul corps politique, en conservant à chacun son indépendance nationale.* Paris: A. Egron.

Savary, Jacques. 1675. *Le Parfait Négociant: ou Instructions pour ce qui regarde le Commerce des Merchandises de France et des Pays Étrangers.* Paris: Chez Jean Guignard.

Say, Jean-Baptiste. 1803. *Traité d'économie politique.* Paris: Deterville. Translation: 2008. *A Treatise on Political Economy.* The Ludwig von Mises Institute. Available on BNF Gallica Digital Library.

———. [1828–1829] 1840. *Cours complet d'économie politique pratique.* Paris: Guillaumin.

Smith, Adam. [1776] 2007. *An Inquiry into the Nature and Causes of the Wealth of Nations.* Hampshire: Harriman House LTD.

Sombart, Werner. [1916] 1927. *Der moderne Kapitalimus.* Munich and Leipzig: Verlag von Duncker & Humblot.

Spence, William. 1807. *Britain Independent of Commerce.* London: T. Cadell and W. Davies, Strand.

Temple, William. 1673. *An Essay Upon the Advancement of the Trade in Ireland.* Dublin.

Tucker, Josiah. 1763. *The Case of Going to War, for the Sake of Procuring, Enlarging or Securing of Trade, Considered in a new light.* London: R. and J. Dodslet.

Vanderlint, Jacob. 1734. *Money Answers All Things, or an Essay to Make Money Sufficiently Plentiful Among all Ranks of People and Increase Our Foreign and Domestic Trade.* London.

Verri, Pietro. [1760] 1804. *Degli Elementi del Commercio.* Milano: Stamperia e Fonderia G.G. Destefanis.

Weber, Max. [1922] 1978. *Economy and Society.* Berkeley: University of California Press.

Wright, Quincy. [1942] 1960. *A Study of War.* Chicago: University of Chicago Press.

# 3

# From Historicizing to the Obsolescence of War

The physiocrats' pacifism derives from the laws of nature, while that of the classical economists was drawn from the laws of economics themselves. The lack of state economic intervention and international free trade formed elements for pacification, each of which was supported both by the physiocrats and the laissez-faire economists and fitted in well with the economic reality of Great Britain. Above all, it suited the needs of the most advanced industrial sectors of steelmaking and cotton textiles, which just then were starting the first industrial revolution and faced no rivals on the world market.

Free trade would instead have penalized the countries starting late in their industrialization—that is all, except Great Britain—because the exchange between agricultural produce and industrial products favoured the latter. The reduction of production costs, thanks to mechanization, took place more fully in industry than in agriculture, which was also conditioned by weather conditions that humans could not alter. The countries which did not engage in industrialization risked remaining in a state of underdevelopment, because their exports of agricultural produce would never earn what was required in terms of industrial goods and technology.



R. Allio, *War in Economic Theories over Time*,
Palgrave Studies in the History of Economic Thought,
https://doi.org/10.1007/978-3-030-39617-6_3

The latecomers to industrialization could not reach the level achieved by Britain based exclusively on private initiative and capital. New technologies and new plant became increasingly expensive. The newborn industries of the backward countries needed a period of protection before they could compete with Great Britain, with state aid and preference in public contracts acting as a sort of incubation that was destined to last until they could stand on their own feet and produce at competitive costs.[1] Backward countries could have to face wars too, in order to succeed in gaining an economic hold.

It is no coincidence that protectionism and state intervention in the economy, even as a temporary measure for the countries left behind in the process of development, were proposed and theorized in the first half of the nineteenth century by the German economist Friedrich List. He was a supporter of laissez-faire theories in his youth, but after spending a few years in the United States and having observed the effects of their economic policy, returned to his own country before its political unification, reflected on the German states' backwardness and industrial potential and converted to protectionism.

List, like other nineteenth-century German economists, opposed the theoretical conception of economics as a precise science as developed by the British classical economists, with an historical reading of the economy. As a human science, economics could not envisage universal laws valid in all cases, but, instead, had to proceed from the observation of reality, seeking to interpret it.

## The Germans, Neomercantilism and Wars

For laissez-faire economists, the cosmopolitanism and pacifism of the economy were not upset by the permanence of states as they did not modify or hinder economic laws, also seeing that national governments must not intervene in the field of the economy, neither domestically, nor in controlling international trade. Since the economic interests of nations are not conflictual, the perspective was for a spontaneous collective opening to the free international market, which would be useful for all states and their populations.

This was not at all the case for Friedrich List, the leading German economist of the early nineteenth century. For him, the theory of free trade was not a universal theory, but rather the egoistic pretence of the strongest who wished to extend to all choices convenient to Great Britain, as if it were a benefit to all. Free trade would really benefit all only in the case that nations had reached a similar level of industrial development. This was not the case in fact.

While taking into account humanity as a whole, on the one hand, and the individual, on the other, classical theory had overlooked the nation, according to List the indispensable intermediary between the two. This led to the creation of *cosmopolitical* economics, as suggested by Quesnay and Smith, which is something abstract and different from political economy or national economics. *Cosmopolitical* economics, as promoted by Quesnay and Smith, holds that all humanity can achieve well-being following the same laws. National economics, on the other hand, seeks to understand how a given nation can concretely develop, thanks to intervention by the government, which acts as the intermediary between the individual and humanity, with its task, in the economic field, of directing domestic production and distribution of wealth in the interest of all citizens, instead of that of the few.

Great Britain alone, thanks to its commercial and industrial supremacy could be interested in free trade because only in a nation like Britain did *cosmopolitical* principles already coincide with political principles.

The classical economists abstracted completely from reality when supporting laissez-faire theses, thereby making their theoretical conclusions inapplicable. After Quesnay and Smith, this "gap", which had separated theory from practice, just continued to widen and List asked himself what use was there for a science that did not set out the way to be followed in practice (List [1841] 1909, Preface).

List, writing after the signing of the Zollverein treaty, but before German political unification, spoke, not by chance, of a "national" system for the economy. The laissez-faire economists did not deal with the theme of war in a particular way just because their theory rejected the reality surrounding them. This saw the presence of national states that were far from always having common interests, but instead followed their own economic preferences, thereby often coming into conflict, consequently leading to the possibility of international hostilities breaking out, and so having to prepare for this outcome. Political economy, a human and historical science, could not attempt to ignore the existence of the problems connected to the possibility of such hostilities.

For the laissez-faire economists, protectionism and trade limitations were the causes of war, while List turned this statement on its head to state that not only does war, military or economic, give rise to protectionism and defensive economic systems, but it also operates as a protectionist system itself during the period of fighting. Thus, in turn, it is much more likely that a war is caused by ruthless international competition than by protectionism.

Peace is certainly needed for the economic development of nations, and history teaches us "that wherever individuals are engaged in wars, the prosperity of mankind is at its lowest stage, and that it increases in the same proportion in which the concord of mankind increases". But the process of pacification, of unification by means of international trade, was then just at the beginning and was frequently interrupted and hindered by wars or by the self-interested measures taken by one or another nation (List [1841] 1909, Preface).

War, for List, was not universally negative from an economic point of view. Conflicts required modern armaments and, since the military sector became increasingly advanced technologically, the ability to produce the armaments needed in a modern war could lead to the transition from a prevalently agricultural economy to a prevalently industrial one. In this case, war is useful in the development of the country. On the contrary: "A peace which throws back into a purely agricultural condition a nation which is fitted to develop a manufacturing power of its own, becomes a curse of it, and is incomparably more injurious to it than a war" (List [1841] 1909, p. 148).

Even though List declared that he did not belong to those who defended useless costs, above all the costs for creating and maintaining large armies, or the costs of war and those caused by the need to service large public debts, he distanced himself from classical theory, which he called "current theory", which held all consumption, for example, the costs of war, which are not immediately reproductive, to be harmful. List foresaw, in fact, the effects that over a century later were defined as the spin-off of war expenditure:

> The equipment of armies, wars, and the debts contracted for these purposes, may, as the example of England teaches, under certain circumstances, very greatly lead to the increase of the productive powers of a nation. Strictly speaking, material wealth may have been consumed unproductively, but this consumption may, nevertheless, stimulate manufactures to extraordinary exertions, and lead to new discoveries and improvements, especially to an increase of productive power. This productive power then becomes a permanent acquisition; it will increase more and more, while the expense of the war is incurred only once for all. (List [1841] 1909, p. 43)

It may therefore be the case that under favourable conditions, such as Britain enjoyed, a nation could earn much more from a war than what it spent on it in terms of the consumption that theoretical economists considered unproductive. If the burden of public debt caused by the war was borne by those who had gained from the war, instead of being paid for by consumer taxes, which reduce the poor to an unbearable limit, then the advantage would be even greater.

List, who nevertheless confirmed the value of peace, did not exclude the possibility and the opportunity of creating an international association to propose universal peace, but this could only happen when nations had reached a sufficiently advanced level, approximately equal, of industrialization and economic strength. In fact, on reaching a high level of economic development, every nation would have an interest in practising free trade and in maintaining peaceful relations with others, as was the case at the time only for Britain.

If, instead, laissez-faire was imposed in too short a time, it could cause damage to the backward countries. The immediate introduction of free trade on an international level in the first half of the nineteenth century would have led to a universal monopoly for British industry and to its international supremacy, along with a parallel decline in all other nations.

Protectionism was therefore indispensable in order to start and consolidate the process of economic growth in "industrializable" countries, those that had access to factors that made them able to industrialize, such as raw materials, sufficient resources and a large civil population, but, for various historical reasons, had fallen behind Great Britain. Customs protection was a temporary historical necessity and offered adequate means to allow for the industrial take off in a country that had the potential, but should not be used to enforce autarkic policies, something List condemned. The mercantilists' error lay precisely in considering protectionism as a permanent system.

Both free trade and international pacification could be achieved in the interest of all concerned with the realization of a sufficiently advanced and similar economic growth that applies to various nations. Economic systems were not, in fact, good or bad in themselves; "it depends on the special circumstances of the nation and on the condition of its industry which of these is the right one to be applied" (List [1841] 1909, p. 248).

As regards the possibility of an imminent war, List did not exclude it, if it should break out in order to create a "German national trade system". On the contrary, he laid down a programme for territorial expansion for the future united Germany that not only referred to the German states, but looked further afield to a confederation that List hoped could include the Netherlands, Belgium and Switzerland, as well as territories lying to the southeast of Germany, in agreement with Hungary. Such a programme had, objectively speaking, a great possibility of being warlike and it was a good idea for the German states to make plans to wage such possible wars.

If the immediate prospect was of conflicts that could involve Germany, and consequently Europe too, List saw in the future even the possibility of an intercontinental war between the United States and Great Britain to gain economic primacy and suggested that the Americans should start to make preparations for it.

Following List, the German historical school of economics made a critique of the abstractions of the classics and then later of the logical-deductive approach of the marginalist economists. This came in two phases: the "first generation" of Wilhelm Roscher, Karl Knies and Bruno Hilderbrand, working in the twenty years between 1840 and 1860, and the "second generation", active at the turn of the nineteenth century, represented mainly by Gustav Schmoller. The concept of personal interest was contrasted to that of the popular spirit. As far as the method was concerned, while for List economics should be based on experience, taking account of history, politics and philosophy, for the historical school the proposal was for a more radical reform of economic science to base it on the historical inductive method.

Even if there were divergences over specific points, the leading students in this school shared some common points: economic policy had to be considered at a national level and, at least temporarily, state intervention and the introduction of protectionism were vital to allow backward nations to start the process of development.

However, as far as war was concerned, the positions of those who considered the question did not seem to be completely homogeneous. An exception could be made for the fact that none of them thought that a situation of long-lasting peace was imminent and, consequently, all of them considered it opportune for nations to maintain a large army, for some for defence and, for the others, for aggression. Wilhelm G. F. Roscher stressed the economic losses that are caused by conflicts following material destruction, but also owing to the fall in production they provoke. This was not only due to the large numbers of young people who are taken off productive work to be enlisted in the army, but also because of the sudden change over from peace to war caused economic upsets and is always accompanied by crises. War also led to a loss of wealth at an international level, because, following the destruction, the winners gained less than what the losers lost.

The defence of states had to be guaranteed by maintaining an adequate army as permanent peace was far off and wars were destined to last a long time. If war preparation was handled with "reasonable" means, it would be productive in an economic sense too and would function as a deterrent to the outbreak of wars.

Karl Knies, Professor of Economics at Freiburg and Heidelberg, put forward some counter theses. He agreed with Roscher in stating that war not only led to destructive action on goods, but also cut production. Nevertheless, wars could favour the economic interests of one country to the detriment of another. In fact, it was not always true that a war was negative for all parties to it. All historical epochs have seen wars provoked by economic interests, firstly for the conquest of land, then for questions of trade and finally for industrial reasons. Even when a war broke out for reasons other than the economy, according to Knies, the participants sought to gain a material advantage all the same, occupying preferably land or imposing war reparations on the vanquished country.

Although there was no question about the fact that the accumulation of capital was higher in periods of peace, war could not be excluded and Knies even accepted the legitimacy of wars of aggression when they served to prevent an attack thought to be planned by enemy (here foreseeing the concept of "preventive wars" brought forward by the United States in this century). In any case, defence has to organized and military expenditure provided for it, and when it clearly did not exceed "what is necessary", could even be the most productive of all state expenditure. Knies considered that the process of civilization of nations was a tendency towards pacification, but felt that long-lasting peace was still far off. War was and remained a frequent event in history and if one wanted to see it as a pathological phenomenon, it had to be considered as chronic (Knies 1853, pp. 85–87; Silberner 1957, pp. 141–142).

Years later, at the turn of the nineteenth century, Gustav Schmoller rejected the pacifism of laissez-faire economists and sociologists, holding that social peace within a state, just like peace between nations, was a pious hope and it was not true that democracy and political freedom contributed to the process of pacification. Nations existed and nothing led anyone to believe that they would have to leave the field free for other political institutions. Nations also had particular interests, which did not coincide and could lead to war. For example, Germany gained power along with Great Britain, the United States and Russia in the late nineteenth century. Schmoller held that, in order to maintain its position as a strong state, Germany had to further increase foreign trade and hold on to its colonies, even conquering others, both tasks that could not be fulfilled out without strengthening the war fleet, a choice that Great Britain would certainly not appreciate.

Sombart devoted much space in his reflections on economics to the theme of conflicts, because he believed that the development of capitalism owed much to the existence of armies and war. The latter was historically the origin of a great increase in demand, an essential condition for the growth of capitalist production (Sombart 1913).

Sombart, who had connections with the German historical school and had analysed Marxist thought, paid particular attention to the years running from the end of the nineteenth century through to the beginning of the twentieth century, the years of the great return to colonialism. The situation that came into being with the depression starting in the 1870s[2] showed that, in reality, when difficulties arose, the economic interest of individual nations prevailed over international ones and that there was no real general common interest, as the laissez-faire economists held. The state followed its own objectives and, in the periods of crisis, fatally entered into conflict with the interest of others.

According to Sombart, in the modern state there was a clash between the principle of power politics, which pushes for conquest and war, and the principle of laissez-faire economics, which foresees international pacification. Furthermore, the protection of the interests of the whole population in foreign policy is often hindered by the special interests of individual groups wielding political influence.

While defining his conception of imperialism,[3] Sombart reconstructs the historical pathway taken in making choices in economic policy carried out by the state from its conception through to the contemporary period. The economic policy of the absolutist state "was univocally determined by its *own* interests [...]. Power and economy, in the age of mercantilism formed an indivisible whole". In foreign policy, the powers fought each other "with all the tricks of cunning and forms of violence", they established colonies "on the basis of the most shameless rapine of people and countries", they established "brutal" customs duties, "brutal maritime law" and, "as the final measure, took recourse to the violence of arms" (Sombart [1916] 1927, p. 60).

Then the wind of history changed direction and the wars fought for the power of the state were overtaken by revolutionary wars fought for an ideal. Therefore, Napoleon, who, according to Sombart, defended the idea of Europe, clashed with the Central and Eastern Powers which opposed him with a different ideal of dynastic legitimacy. Starting at the end of the Napoleonic Wars yet another ideal began to spread, that of laissez-faire, so that towards the mid-nineteenth century:

> The foreign policy of European states began to turn toward free trade and thus toward the idea of peaceful community of trade for all peoples, seen as being dissolved into atoms and thus, more precisely, for all individuals and all economies, that had to unite in a social cosmos on the unhindered road of trade, the final outcome of such social Newtonism being the idea of free trade. (Sombart [1916] 1927, p. 61)

Sombart, like List, thought that the free trade movement did not start in Britain by chance or as a mere theoretical consideration. In fact, having become "the workshop of the world" following industrial expansion, it had to sell abroad the products in excess of domestic demand. Other empires, however, Russia in the lead, did not share the same interest, and consequently the laissez-faire movement did not gain an overall recognition. The depression towards the end of the nineteenth century led to an "about turn" and foreign policy returned to the leading motive of the interest of the state.

In the second half of the 1880s, Sombart's reading of the predominant voluntaristic and colonial policy saw it as a return to mercantilism, but with a fundamental difference, that nineteenth-century neomercantilism was determined by the interest of capitalism and no longer by the interest of the state, as had been the case in the mercantile epoch. "Previously the state had directed the economy, while later the economy directed the state" (Sombart [1916] 1927, p. 63).

Therefore, Europe, with the exception of Great Britain, returned to protectionism. Expansionist tendencies regained strength in the major powers, aiming to stretch their area of economic influence beyond the borders of the homeland. The militarization of states grew with the arms race and the increase in defence spending.

Alfred Weber, the German economist and sociologist, brother of Max, also (1925) considered the militarization of states, rearmament and the increase of spending on the army. Sombart took up these considerations and spoke of the development of militarism, which advanced hand in hand with capitalist growth and which, like capitalism, followed its own laws of development based on many aspects, however, that were coming to the fore independently of economic development. So, there came about a "realignment of capitalism, which had become a political power, with the growing militarization of the state which, even if springing from its own natural tendencies, produced the same effect" (Sombart [1916] 1927, p. 66).

When considering the causes that generated the expansionist interventionism, Sombart thought that Lenin's thesis was wrong, or at least unilateral, seeing that imperialism was also present in other zones, such as Russia and Japan, which had not given rise to significant forms of finance capitalism between the end of the nineteenth and the beginning of the twentieth centuries. In Switzerland, however, a country with highly developed finance capital, there was no trace of imperialist tendencies. According to Sombart, the causes of "such a great" phenomenon could not be due solely to class interests and economic motives. He thought that many other motives contributed to determining imperialism. They included political factors, which expressed themselves in the will to power and the expansionist tendencies of the state, military motives, and here Sombart quotes Schumpeter who thought that the military machine came with its own automatic expansionist tendency (Sombart [1916] 1927, p. 68), then too, nationalist motives, demographic motives and, lastly *capitalist* motives.

The problem, according to Sombart, did not lie in understanding what role the economy played in the development of imperialism, but, vice versa, research should be made into the importance of imperialism for the economy.

The principle economic manifestation of imperialism was then neomercantilism, which had provided capitalism with the instruments of the strong state. The laissez-faire ideas, the free competition of the state economies had been shown to be "inadequate for capitalism" in the phase of depression. If capitalism was able to establish on the domestic market the instruments of power required to follow its own interests with its own forces alone, in relations abroad it could not do without "the instruments of power of the sovereign state" (Sombart [1916] 1927, p. 69), that is to say, public intervention. In the case of Germany, protectionism had allowed the industry of semi-finished goods to grow; it had accelerated the formation of cartels and had contributed to the expansion of the market with export subsidies. These were all positive factors for capitalism.

Further still, the expansionist policy of neo-imperialism was useful for capitalist development, not so much because raw materials were to be found in the colonies at a low cost, nor because the conquered territories could absorb finished products lacking a domestic market, but rather because they were places to invest the excess capital of the homeland. Here Sombart contradicts his previous criticism of Lenin.

Militarism too, fed by the imperialist policy, came to be useful to capitalism because the growing demand for military supplies favoured armaments production and, higher up the supply chain, it supported "the most important sector of capitalism, heavy industry" (Sombart [1916] 1927, p. 71).

In 1915, during the First World War, Sombart published a warmongering pamphlet awash with patriotic spirit entitled *Merchants and Heroes* in which the English were depicted as a people of shopkeepers, unlike the German philosophers and heroes. The war would seize the whip from the hand of the Anglo-Saxon race which for centuries, with the help the Dutch and the Jews, had held Germany in the corrupting vice of great international finance, keeping it in a state of economic inferiority (Sombart 1915; cfr. Boldizzoni 2008, p. 138).

## War as Atavism

The few laissez-faire economists who expressed an opinion on war after the First World War judged that the use of violence ran against the development of the economy. Mises held that, through "the Liberal Social Philosophy, the human mind becomes aware of the overcoming of the principle of violence by the principle of the peace", because warlike conflicts are damaging not just to the losers, but also to the victors. Society was born from works of peace and its task was to maintain it. It was the economy, not war, that created widespread well-being. Peace was therefore "*the* social theory of Liberalism" because maintaining the protection of property, as laissez-faire economists did, was the same thing as rejecting war (Mises 1962, pp. 69–70). The only war that could be allowed by liberal society was war of self-defence.

Mises always closely followed the laissez-faire letter when he forcefully criticized the positive judgement on the state war economy expressed by Otto Neurath ([1919] 2004), who, in the immediate post-war period, proposed to extend the model to peacetime too. Neurath, a philosopher, sociologist and economist, was a member of the Vienna Circle, holding a left-wing position. He was one of the first to analyse the particular characteristics of the war economy, which he defined as a planned economy aimed at satisfying the needs of citizens and carried out by coercive means only because of the particular circumstances in which it was tried. During the brief period of the Munich Soviet Republic in 1919, Neurath was employed in fact in the reconversion of the war economy to a planned peacetime economy.

Mises opposed this hypothesis of economic planning and instead maintained that major state intervention in the economy, accompanied by a reduction in the role of money, was in reality a major step backwards towards primitive forms of social and economic organization based on the use of force.

However, Joseph Schumpeter was the man who above all sought to save the idea of the pacific vocation of industrial societies.[4] He presented the First World War as a tragic aberration due to the presence, above all in Germany and Japan (two countries which upset the equilibria of the twentieth century), of cultural and institutional elements typical of the feudal aristocratic spirit of the ruling class tied to the values of militarism.

Schumpeter repeated things widely voiced beforehand: war and imperialism were leftovers of preindustrial society, in which the nobility gained wealth and power through the use of violence and the right to take booty. In a society where productive techniques did not permit rapid and major increases of wealth, the most common way to increase the holdings and income of the individual or the state was to resort to aggression and violence.

> Imperialism thus is atavistic in character. It falls into that large group of surviving features from earlier ages that play such an important part in every concrete social situation. In other words, it is an element that stems from the living conditions, not of the present, but of the past – or, put in terms of the economic interpretation of history - from the

> past rather than present relations of production. It is an atavism in the social structure, in individual, psychological habits of emotional reaction. (Schumpeter 1951a, pp. 84–85)

For Schumpeter, as for the classical economists, the rise of the bourgeoisie to become the ruling class should have led to peace, since this class was able to gain and increase wealth through technological improvement and the rationalization of production and distribution systems.

> Unlike the class of feudal lords, the commercial and industrial bourgeoisie rose by business success. Bourgeois society has been cast in a purely economic mold: its foundations, beams and beacons are all made of economic material. The building faces toward the economic side of life. Prizes and penalties are measured in pecuniary terms. Going up and going down means making and losing money. (Schumpeter 1947, p. 73)

The bourgeoisie, in its pursuit of wealth, substituted rationality for force and brutality; bourgeois, capitalist society is quite antiheroic "fundamentally pacifist and inclined to insist on the application of the moral precepts of private life to international relations" (Schumpeter 1947, p. 128). In a purely capitalist world, the energy once used in war would instead be employed in all kinds of productive work. Wars of conquest and adventurism in foreign policy were a dangerous distraction, destructive of life and a diversion from the habitual and true tasks of society.

From the individual's point of view, as well as of collective behaviour, war was therefore tied to living conditions of the past, to old modes of production, to superseded social structures, to psychological conditions and atavistic emotional reactions. Consequently, it was not true, as Marxists held, that capitalism tended to develop forms of imperialism and, all the less so, that imperialism was the necessary and last stage of capitalism: "*It is a basic fallacy to describe imperialism as a necessary phase of capitalism, or even to speak of the development of capitalism into imperialism*" (Schumpeter 1947, p. 118. Italics in the text).

The fact that militarism was the "declining residue of preceding epochs" (cfr. Mann 1980, p. 27), before Schumpeter expounded it, had already been explicitly or implicitly stated by the majority of classical economists and by Molinari, Pareto and positivist sociologists too. Hobson himself, at the end of his work on imperialism, had stressed how it was "a depraved choice" imposed on nations by selfish interest that had survived from distant centuries of the struggle for survival, interests that in contemporary society called on greed and gain and was based on violent imposition (Hobson 1988, p. 368).

However, the world war was too strong a rebuttal of the utopic expectations of the laissez-faire economists and required a new effort for an explanation. It was just Schumpeter who tried to do this by holding that the warlike violence, even in an extreme form as in the world war, erupted because *subsidiary factors* were at work in contemporary society that enabled irrational mindsets and obsolete structures aimed at violence to survive. Throughout human history, "aimless" tendencies had played a major part in the spread of violence, playing no heed to utilitarian limits.

Schumpeter nevertheless specified that the "functional aggressive need" and the "will to war" in present-day society did not depend on mere whims or instincts, but on deep rooting and consolidation over time of vital needs that, in the distant past, forced men to be violent in order to avoid extinction. In other words, contemporary aggressive tendencies were none other than the leftovers of the struggle for survival, as Hobson had previously said. These psychological mindsets were perceived and remained active long after having lost any significance and usefulness in safeguarding life. According to Schumpeter, both the interests of domestic policies of the ruling classes and the influence of all those who thought that could gain individual advantage at a financial or social level from a warlike policy, conspired to keep these tendencies in operation in an industrial capitalist society. They consequently bolstered forms of atavism as well as the inclination to go to war. Elements "of a psychological nature could be added to these factors, so that many, perhaps most wars have been waged without "adequate 'reason' not so much from the moral viewpoint as from that of reasoned and reasonable interest" (Schumpeter 1951a, p. 83).

The capitalist world should not offer any space for imperialist impulses, but this did not mean that contemporary society would not see thrusts towards territorial expansion, because particular interests remained which allied with an "imperialistic tendency flowing from non-capitalistic sources, to use them, to make them serve as pretexts, to rationalize them, to point the way toward action on account of them" (Schumpeter 1951b, p. 97).

Schumpeter repeated that imperialism would disappear bit by bit, because industrial workers too, and not just the bourgeoisie, were against violence and would succeed together in conquering the residual resistance of the bodies in power. Imperialism would gradually disappear because the vital needs that had generated it "have passed away" (Schumpeter 1947, p. 72). It was true that every fresh war tended to reactivate imperialism, but Schumpeter did not doubt that it would disappear as a structural element, because the structure it stood on goes into a decline. All this simply takes us back to what the laissez-faire economists had previously stated.

Schumpeter admitted that the industrialists in the armaments sector of capitalist society were interested in war and that, if the war lasted long enough, other sectors too could gain an economic advantage. Meanwhile, the economy in general would be impoverished due to the excesses of consumption and the destruction of wartime, but the profits of capitalists as a social class did not constitute a valid motive to start conflicts (Schumpeter [1927] 1951b, p. 98). Of course, the industrialists in the armaments sector were influential, but they formed a minority. According to Schumpeter, there were few researchers who would go as far as stating that this element alone was sufficient to swing public opinion in the capitalist world towards imperialist tendencies (Schumpeter 1951b, pp. 98–99).

Capitalists and leading entrepreneurs therefore did not have to be nationalistic, in theory, but nationalism still prevailed. Schumpeter laid the blame on the intellectuals who spread it, on the bourgeoisie that bowed down before the power of the political aristocracy and on the monopolistic exporters who pushed for colonial conquests.

The same held true for militarism and imperialism. A nation could be defined as militarist not when it had a large army, but rather when the high command exercised a significant political influence over society in general. In industrial society, it was not the capitalists who wanted imperialism, but the demagogues, the politicians interested in it, the military and groups of intellectuals who manage to gain the upper hand over the pacific behaviour of the bourgeoisie.

What separates Schumpeter's thought from that of the classical economists also forms the weakest point in his theory. When he spoke of imperialism, Schumpeter did not mean the behaviour of the state, which, even through acts of war, pursued its own real political and economic interests. Imperialism for him meant, hegemony, domination, the desire for supremacy and expansion for expansion's sake. This behaviour was a leftover of the absolute monarchies and past living conditions, which contemporaneously would regard the social structure, the psychic habits and emotive reactions of individuals, while, from an economic point of view, would refer back to superseded modes of production. Expansionism, except when it is the expression of a society of warriors, for Schumpeter was not imperialism. Consequently, since capitalism was not based on a social class of warriors, it was not imperialist.

Schumpeter's theory has drawn a lot of criticism. In particular, Murray Green (1952) identified the weakest point precisely in the rather loose definition of imperialism, considered as a mindset without an aim, on the part of the state, of unlimited expansion. This aim would derive from the social structure of the war-making class, which would seek expansion in order simply to expand, war in order to fight and domination for its own sake.

When considering the boasted capitalist rationalism against war, it was yet again Murray Green who reminded Schumpeter that it was precisely capitalism that invented racism and chauvinism, dreamt up "the white man's burden", spoke of the "manifest destiny" and proposed the doctrine of national interest. Greene concluded that Schumpeter's attempt to propose a theory that was valid for Ancient Egypt as well as feudal France or for twentieth-century capitalism came down in fact to a distortion of the nature of capitalism and was an ahistorical abstraction.

Perry Anderson, writing back in the 1970s about the "absolutist" state over time and in space, held that warlike conflicts did not correspond with capitalist rationality, but instead represented the emphasis of the medieval function of the war, when this was the profession and destiny of the ruling class, which managed to use war to maximize wealth.

War lost the role and was no longer economically rational in societies that had developed the industrial mode of production and were characterized by a higher rate of accumulation of capital and the rapid change of economic institutions and which could, therefore, produce new wealth. The permanence of international armed conflict was therefore one of the visible signs of the permanence of an absolutist environment (Anderson 1974, pp. 31–33). The long list of conflicts in the twentieth century seemed for Anderson to lie outside the needs of capital, even if it too in the end contributed to it (Anderson 1974, p. 33). Michael Mann commented acidly on this when he said that after 95 million deaths in two world wars, fought almost exclusively by mature capitalist countries, Andersen's interpretation seemed very difficult to accept (Mann 1980, p. 27; 1988, p. 126).

From a political point of view, in 1955 Quincy Wright noted in his studies into international relations that, on the other hand, no civilization had ever arrived at the state of the obsolescence of war and how, when everything was taken into consideration, the democracies did not appear any bit less warlike than the autocracies (Wright 1955, pp. 154, 841).

## Notes

1. The basic text dealing with the problems backward countries must face and the "advantages" they have is still A. Gerschenkron. *Economic Backwardness in Historical Perspective*, 1962.
2. Caused by monetary problems and the unequal distribution of income, the depression was seen mainly as the stagnation of prices, the slow adsorption of supply and the lack of profitable investment opportunities for capital.
3. Sombart used the term imperialism exclusively when referring to state policy, while he considered out of place to speak of cultural imperialism, social imperialism, etc.
4. "True, pacifism as a matter of principle had existed before, though only among a few small religious sects. But modern pacifism, in its political foundations if not its derivation, is unquestionably a phenomenon of the

capitalistic world". "Capitalism is by nature anti-imperialist" (Schumpeter 1951b, pp. 92, 96).

## References


Anderson, Perry. 1974. *Lineages of the Absolutist State.* London: New Left Books.

Boldizzoni, Francesco. 2008. *Means and Ends. The Idea of Capitals in the West, 1500–1970.* Houndmills, Basingstoke, and Hampshire: Palgrave Macmillan.

Gershenkron, Alexander. 1962. *Economic Backwardness in Historical Perspective.* Cambridge: Belknap Press of Harvard University Press.

Green, Murray. 1952. "Schumpeter's Imperialism. A Critical Note." *Social Research* XIX (4) (December): 453–463.

Hobson, John Atkinson. 1988. *Imperialism. A Study.* London: Unwin Hyman Ltd.

Knies, Karl. 1853. *Die politische Ökonomie vom Standpunkte der geschichtlichen Methode.* Brunswick.

List, Friedrich. [1841] 1909. *The National System of Political Economy.* London: Longmans, Green and C.

Mann, Michael. 1980. "State and Society 1130–1815: An Analysis of English State Finances." In *Political Power and Social Theory,* edited by Maurice Zeitlin, vol. 1. Connecticut: Jai Press.

———. 1988. *State, War and Capitalism.* Oxford: Basil Blackwell.

Mises, Ludwig von. 1962. *Socialism. An Economic and Sociological Analysis.* New Haven: Yale University Press.

Neurath, Otto. [1919] 2004. *Economic Writings: Selections 1904–1945.* Dordrecht: Kluver.

Schumpeter, Joseph Alois. 1947. *Capitalism, Socialism and Democracy.* London: George Allen & Unwin Ltd.

———. [1927] 1951a. *Imperialism and Social Classes.* New York: Augustus M. Kelley, Inc.

———. [1927] 1951b. "The Sociology of Imperialism." In *Imperialism and Social Classes,* 3–130. New York: Augustus M. Kelley, Inc.

Silberner, Edmund. 1957. *La guerre et la paix dans l'histoire des doctrines économiques.* Paris: Sirey.

Sombart, Werner. 1913. *Der Bourgeois: zur Geistesgeschichte des modernen Wirtschaftsmenschen München*. Leipzig: Duncker & Humblot.

———. 1915. *Händler und Helden: patriotische Besinnungen*. München: Dunker & Humlot.

———. [1916] 1927. *Der moderne Kapitalimus*. Munich and Leipzig: Verlag von Duncker & Humblot.

Weber, Alfred. 1925. *Die Krise des Modernen Staatsgedankens in Europa*. Struttgar: Deutsche Verlans- Anstalt.

Wright, Quincy. 1955. *The Study of International Relations*. New York: Appleton-Century-Croft Inc.

# 4

# Imperialism

Marxism, unlike the classical and neoclassical schools of economics, has its own theory of war and has studied its causes. Nevertheless, Marx did not write a great deal on war and *Capital* did not consider militarism. For Marx, the source of war was not found in a form of government and it was not the *Ancien Régime* aristocracy that caused it, but rather a mode of production, the capitalist mode of production. Class antagonisms lay at the origin of the struggles between the bourgeoisie and the proletariat and led to socialism, which would eliminate the bourgeois state and with it war too. The problem of peace was thus solved automatically.[1]

In his contribution to a *Critique of Political Economy*, Marx placed the emergence of capitalism in the post-renaissance period when the national states were coming into being and were arming themselves as they foresaw wars to establish their frontiers. Capitalist production therefore was born more for war than for peaceful production.

Engels showed frequent interest in the questions of military science and the art of war and also wrote much on violence. All the articles and letters published in the complete works on the subject were collected in a single volume published in France in 1947.


R. Allio, *War in Economic Theories over Time*,
Palgrave Studies in the History of Economic Thought,
https://doi.org/10.1007/978-3-030-39617-6_4

Both Marx and Engels made frequent reference to war and colonialist episodes taking place in their lifetimes both in articles and their private correspondence. They analysed above all the effects of the First Opium War and British colonialism in India and Ireland. The overall judgement of the episodes regards the intrinsic connection between capitalism and colonialism and the predatory nature of colonialization hiding behind the mask of a presumed civilizing function.

The fact that capitalism bore violence that was not only internal, but also external, was the object of analysis by Marxists above all at the end of the nineteenth century and during the first decades of the twentieth century. This period was marked by the creation of the great industrial trusts, of massive rearmament, of neocolonialism and, lastly, the outbreak of the First World War. The literature on the question is too great and too well known for anything of value to be added to it. I will therefore limit this chapter to indicating the main arguments of the various conclusions reached by those who made reference to Marx in applying his thought to the reality of late nineteenth-century European imperialism and the First World War. However, before dealing with the Marxists' analysis, we have to consider the earlier study by Hobson.

## Hobson

As is well known, at the beginning of the twentieth century, Hobson provided a definition and offered an interpretation of warlike imperialism, which provided the starting point both for the Leninist reading of it and the critical laissez-faire one. Hobson noted that Comte's forecast had not come true and that bankers and industrialists seventy years after the publication of *Cours de philosophie positive* still wanted war. Just this little caste of international financiers, operating through the sold-out press, which wished to make exceptional profits and guarantee a low-cost workforce, was blamed by Hobson for the outbreak of the Second Anglo-Boer War (1900, p. 229).

The most lucrative sectors in South Africa, at the time, were the gold mines, the railways, the dynamite monopoly and the alcohol trade. In 1899, on the eve of the war, Hobson studied the amount of profit that entrepreneurs in these sectors could make while he was in the country acting as the correspondent for the "Manchester Guardian". The profits attracted hordes of European investors. This study was Hobson's first attempt to make the connection between imperialist policy and the interests of certain productive sectors. However, wars did not break out only to conquer territories to invest capital profitably. Even if this factor remained the most important in explaining imperialism, the colonies were also useful for disposing of industrial surpluses that entrepreneurs could not sell on the domestic market.

This heterodox, even heretical, liberal therefore attacked British colonial policy at the beginning of the twentieth century, with the Anglo-Boer War in full swing. He accused it of being harmful for the taxpayer and for the economy in general, even though it turned out to be rather profitable for some productive sectors that had managed to impose their interest on the groups in power and used national resources for their own ends. These interventions were undertaken, camouflaged under a noble civilizing intent, not only by Great Britain. Hobson in fact warned that

> The adventurous enthusiasm of President Theodore Roosevelt and his 'manifest destiny' and 'mission of civilization' party must not deceive us. It was Messrs, Rockefeller, Pierpont Morgan, and their associates who needed Imperialism and who fastened it upon the shoulders of the great Republic of the West. (Hobson [1902] 1988, p. 77)[2]

The new imperialism, for Hobson, was connected to the structural transformation underway in capitalist production that was leading to the creation of great monopolies. Another characteristic of neo-imperialism was competition arising between rival empires, while past imperial history saw only one dominant empire taking the stage. From a political point of view, neo-imperialism was the territorial expansion of the autocracy.

Imperialism caused wars and expanded militarism, which in turn greatly increased expenditure on armaments at the expense of projects for social services: public education, old-age pensions, improvement to town and country planning and civil housing stock. On the other hand, it was simply not true that constructing a strong defence system served in preserving peace: *si vis pacem para bellum* presupposed that a real and constant hostility over interests existed between the states. This was not the case. The hostility was between cliques, which "usurping the authority and the voice of the people, use the public resources to push their private interests, and spend the blood and money of the people in this vast and disastrous military game, feigning national antagonisms which have no basis in reality" (Hobson [1902] 1988, p. 127).

Economically speaking, neo-imperialism was a bad deal for Great Britain because at "enormous expense it had procured a small, bad, unsafe increase of markets, and has jeopardized the entire wealth of the nations" (Hobson [1902] 1988, p. 46). Therefore, the question was why Great Britain had embarked on such an unreasonable policy. Hobson's reply was that expansionism had been a good business for some social classes and certain productive activities that were able to influence politics. "The vast expenditure on armaments, the costly wars, the grave risks and embarrassments of foreign policy, the checks upon political and social reforms within Great Britain, though fraught with great injury to the nation, have served well the present business interests of certain industries and professions" (Hobson [1902] 1988, p. 46).

In the end, the financial result of imperialism meant

> a great expenditure of public money upon ships, guns, military and naval equipment and stores, growing and productive of enormous profits when a war, or an alarm of war, occurs; new public loans and important fluctuations in the home and foreign Bourses; more posts for soldiers and sailors and in the diplomatic and consular services; improvement of foreign investments by the substitution of the British flag for a foreign flag, acquisition of markets for certain classes of export, and some protection and assistance for British trades in these manufactures; employment for engineers, missionaries, speculative, miners, ranchers and other emigrants. (Hobson [1902] 1988, p. 48)

At the beginning of the twentieth century, in peacetime, Great Britain spent at least sixty million pounds on armaments, "when hungry mouths, ill-clad backs, ill-furnished house indicate countless unsatisfied material wants among our population" and statistics showed that over a quarter of the English urban population lived "at a standard which is below bare physical efficiency" (Hobson [1902] 1988, p. 86).

Those who gained directly from military contracts were obviously in favour of imperialist policy, but others too upheld it. Hobson listed among them industrial producers in goods that could have been exported to the conquered territories; the capitalists intended to go and build infrastructure in these territories because they lacked profitable investment in the domestic market; shipbuilders and shipping lines; the state bureaucracy and, naturally, the armed forces. All in all, the British colonies were still like James Mill had described them in his own time "a vast system of outdoor relief for the upper classes" (Hobson [1902] 1988, p. 51).

Military, clerical, academic and state bureaucratic circles saw to furnishing sentimentalism and ideology for the concrete and solid financial reasons and to creating a tendency favourable to imperialism in the world of culture. Pressure on the political world was instead exercised directly by economic interest and so it was useless "to attack Imperialism or Militarism as political expedients or policies unless the axe is laid at the economic root of the tree and the classes for whose interest Imperialism works are shorn of the surplus revenue which seeks this outlet" (Hobson [1902] 1988, p. 93).

Productive capacity had grown much faster than wages and so too consumption in Great Britain, as in the United States, Germany and all those countries that one after another engaged in industrial development. Unlike what the "old economic theory" held, the rise in production was unable to promote a corresponding growth in consumption thanks to the fall in prices.

Nevertheless, as Hobson commented, despite the failure of the law of supply and demand, "It is not inherent in the nature of things that we should spend our natural resources on militarism, war, and risky, unscrupulous diplomacy, in order to find markets for our goods and surplus capital" (Hobson [1902] 1988, p. 86).

The overproduction, which led to cartels in the United States and caused difficulties in finding outlets for all industrialized countries, would not have been a problem if income distribution had been better and more equal, that is, if the level of domestic consumption had increased at the same rate as productive capacity. If this had been the case, domestic consumers would have seen their needs transformed into demand and there would have been no need to "force" the way into foreign markets, find colonies or press for war.

## The Marxists

The end of the nineteenth century, when overproduction in various sectors became evident and caused the reaction of the creation of industrial concentrations and monopolies, a return took place to protectionism and colonial conquests, while capital exports and the international financial market expanded rapidly. Liberal-style economic thought faced these new facts by withdrawing from reality and seeking refuge in the theories of the "marginalist revolution". Marxists instead concentrated on the analysis of imperialism and colonialism and the economic reasons for them rather than on the consequences of war. Hilferding and Rosa Luxemburg were the first to grasp the nature of the change, followed by interventions by Kautsky, Bukharin and Lenin.

Even though Hilferding was unable to provide a complete methodological systemization of his research, he did supply Lenin and Bukharin with the foundations on which to build the theory of imperialism through the definition of finance capital. Hilferding examined the productive, distributive and credit systems at the end of the nineteenth century and then summed up the characteristics of the new form of capitalism that had forcefully emerged in Germany in that period. This form, he thought, had two main characteristics. It substituted the system which was tendentially free trade and competition-based up to that on with a new monopolistic system both in production and trade, while at the same time it transformed banks from commercial credit institutes into financial institutes, which intervened directly in the industrial sectors they financed. This new situation developed in the immediate period into "the concentration of economic and political power in the hands of the capitalist oligarchy", producing "the climax of the dictatorship of the magnates of Capital".

In fact, it makes

> the dictatorship of the capitalist lords of one country increasingly incompatible with the capitalist interests of the other country, and the internal domination of capital increasingly irreconcilable with the interest of the masses of people, exploited by finance capital but also summoned into battle against it. In the violent clash of these hostile interests the dictatorship of the magnates of capital will finally be transformed into the dictatorship of the proletariat. (Hilferding [1910] 1981, p. 370)

The perspective traced by Rosa Luxemburg was different. She thought that the neo-imperialism of the late nineteenth century was due to the need for capitalism to find external outlets to counteract the fall in the rate of profit. The economic system to continue to survive required an external market, a non-capitalist world alongside the capitalist one. This need had already been mentioned by Sismondi in *Nouveaux principes d'économie politique* when he held that "with the concentration of fortunes [...] the domestic market is continuously shrinking, and increasingly industry is forced to find new foreign markets" (Sismondi [1819] 1927, p. 361).[3] Imperialism is just the will of a state to ensure, either by force or by threat, privileged outlets for its goods, its capital and emigration through the colonization of non-capitalist areas and economies. These colonial wars were ineluctable and were due to the contradictions of capitalism itself, which caused crises of overproduction.

Rosa Luxemburg focused her attention on the problems of underconsumption. She insisted that the capitalist economy had to conquer external territories for the capitalist mode of production in order to offload products that could not be sold on the domestic market. There the potential consumers, the proletarians, whose wages were kept at a level of subsistence following the subtraction of surplus value by the capitalists, did not have sufficient purchasing power to absorb all domestic production.

Since non-capitalist domestic markets, represented by peasants and craftsmen, were in turn unable to absorb all surplus production, the capitalist states clashed in war over the possession of colonies they wished to influence in order to allow entrepreneurs and capitalists to sell goods and invest savings in excess on the domestic market.

Imperialism was therefore a specific mode of accumulation and constituted the last phase of capitalist development, the one leading to its collapse. In fact, if capitalism needed to establish exchange with pre-capitalist economies, it was precisely through this exchange that these markets would be destroyed as such by being modernized and substituted. This would lead over time to the proletarianization of workers in the colonies too, the point being reached when there would no longer be any economies different from capitalism and any further expansion would be precluded. In this final phase, economic and political upheaval and the class struggle would lead to the proletarian revolution and socialism.

Rosa Luxemburg dedicated the last chapter of *The Accumulation of Capital* (1913) to militarism, defining it as "a means of the highest quality" to realize surplus value, the very nature of capitalism, having been present in every historical phase of the process of accumulation from its dawn, when European capitalism began the colonial conquest of the New World, destroying the local social communities, imposing on them the western way to produce and sell products. Thenceforth war expenditure had always been an effective way to realize surplus value because it was largely financed by indirect taxes and protective customs duties which caused prices to rise and were a burden on the working class. This means that a part of the purchasing power of workers, which could otherwise have been used to purchase means of subsistence, was transferred to the state. The state used this money to increase the demand for armaments. At first sight, this could appear simply as the shifting of demand from one sector to another, but at the same time, there was the transfer of variable capital to state investment, causing a fall in real wages. The use of the tax receipts paid by workers in the production of armaments allowed capital a still further possibility for accumulation. In practice, militarism worked "on the basis of indirect taxation by securing the maintenance of the standing army – the organ of the rule of capital – and by establishing an unparalleled sphere of accumulation: it fulfills both of these functions at the expense of the regular living conditions of the working class" (Luxemburg [1913] 2015, vol. II, p. 339). Therefore, a synergy existed between the overseas activity of militarism aimed at colonial conquest and the internal function addressed to the increase in military expenditure.

> The more forcefully capital uses militarism in order to assimilate the means of production and labour-power of non-capitalist countries and societies through foreign and colonial policy, the more powerfully the same militarism works progressively to wrest purchasing power at home, in the capitalist countries themselves, from the non-capitalist strata – i.e. from those engaged in simple commodity production, and from the working class. (Luxemburg [1913] 2015, vol. II, p. 341)

Thus, militarism increasingly deprives simple commodity production of its productive forces and increasingly reduces the standard of living of workers to provide an impulse instead "at the expense of both" to the accumulation of capital. "From both sides, however, once a certain level has been reached, the conditions for the accumulation of capital turn into the condition for its demise" (Luxemburg [1913] 2015, vol. II, p. 341).

For Lenin too, imperialism was the last phase of capitalism, but the reasons that determined this were different from the ones identified by Rosa Luxemburg. In *Imperialism, the Highest Stage of Capitalism*, Lenin considered the basic economic characteristics of imperialism, going back to Hilferding and Hobson, while criticizing Kautsky at the same time. He added a third characteristic to finance capital to the two identified by Hilferding: the high concentration of production and capital that leads to the formation of monopolies.

Whereas Hobson used his description of the imperialist movement at the end of the nineteenth century to show how some minorities sought to realize super-profits at the expense of the general interest through colonial conquest, Lenin, who, nevertheless, considered Hobson's book as "fundamental" held that imperialism cannot be blamed on minorities, but well expresses the needs of capitalism to reach the final stage of its evolution, the monopoly phase. Free competition determines the concentration of production and this "at a certain level of development" leads to monopoly. The great monopolies could draw an advantage form the colonies as it allowed them to extend their control over resources and markets in the conquered territories.

Kautsky had written several articles and essays on imperialism and war between the end of the nineteenth century and the First World War to repropose or to rectify, under the pressure of events, his theory that counterposed the block formed of reactionary classes and finance capital that was favourable to imperialism, and industrial capital which was instead favourable to free trade. Kautsky's thought was often confused. His 1907 pamphlet entitled *Socialism and Colonial Policy* held that the colonies of the late nineteenth century were "escapes" for capitalism to deal with the crisis of overproduction, while the arms race served "to get rid of the surplus, products…by wasting them" (Kautsky 1907, Chapter VI "New Style Exploitation Colonies" in Marxists Internet Archive—MIA). However, during the First World War he saw another possibility in the imperialisms of the entire world of joining together in great international combinations, instead of fighting each other, creating an ultraimperialism or superimperialism, a holy alliance of imperialists that would put an end to the arms race and the colonial policy, causing war to end within the capitalist regime. At the same time, however, the collective exploitation of the world of labour would favour the creation of conditions for the transition to socialism.

Kautsky's idea was rejected both by Bukharin (1918), who thought it was unrealistic, and by Lenin who called it a "silly story", both wrong and non-Marxist, dreamt up by a petit bourgeois who wanted to hide from reality and console the exploited masses with the hope that permanent peace could be realized under the capitalist regime.

Imperialism was unavoidable for capitalism, but for Lenin and Bukharin it was a phase, the last. Karl Kautsky instead thought that imperialism was a policy of capitalism, more precisely, the "preferred" policy of finance capital. If what was meant by imperialism was cartels, protective customs duties, domination by financiers and colonial policy, the need for capitalism of imperialism boiled down for Kautsky to a "banal tautology" (Kautsky 1914).[4]

Another cause for disagreement hinged on Kautsky's definition according to which "imperialism is the product of highly developed industrial capitalism. It consists in the striving of every industrial capitalist nation to bring under its control or to annex all large areas of *agrarian* [Kautsky's italics] land, irrespective of what nations inhabit it" (Kautsky 1914; Lenin [1916] 1975, chap. VII, p. 85). Lenin agreed that imperialism in general was "a tendency to violence and reaction" but observed that this was not characteristic of industrial capital, but instead of finance capitalism (as Kautsky would also sustain later on) and was not aimed specifically at the conquest of agricultural land, but any type of territory and assumed "an essential feature of imperialism: the rivalry between several great powers in the striving for hegemony, i.e. for the conquest of territory, not so much directly for themselves as to weaken the adversary and undermine *his* hegemony" (Lenin [1916] 1975, Cap VII, p. 85).

Bukharin, like Lenin, started from Hilferding's analysis of finance capital and concluded, like Lenin, that world competition between capitalist monopolies would logically lead to imperialism and war, but while Lenin adopted the model of private national capitalism analysed by Hilferding, Bukharin developed a new one: "state capitalism" or even "state capitalist trust" derived from what was happening in Europe at the time, that was the return to highly state enterprise managed economic policies. Bukharin imagined that industrial and finance capital would have merged with that of the state and would have been incorporated to such a point that the national economy would be transformed into a state capitalist trust. The war would be the lever for this structural transformation, but the effects would have been permanent.

The future therefore belonged to state "organized capitalism" that would be able to create a strong apparatus able both to avoid the anarchy of private market economies and to supersede the internal contradictions of capitalism identified by Marx. This change would not avoid the collapse of capitalism and the transition to socialism, which however would be derived from wars caused by competition between imperialist states.

Imperialism too would have become a state initiative and imperialist competition would have taken place simultaneously at an economic, political and military level. The imperialist state would have demonstrated predatory attitudes, because the aggressive behaviour would have shifted to an international level, while the crises would no longer have been the cyclical trend of national economies, but more the result of conflicts between states. The war would have been the final form of these crises and form the context in which the proletarian revolution would have developed (cfr. Shaw 1984, pp. 57–58).

As Martin Shaw observed, the reflections on war on the part of Bukharin and Lenin led to completely different outcomes. Lenin, who stressed the contradictory nature of war, held that it led to creating the conditions for the revolution, while Bukharin, who saw capitalism linked to militarism, considered the war as the prime cause of the crisis of the system, replacing the role of overproduction, as was the case in the Marxist tradition (Shaw 1984). Shaw considers imperialism as due to the general propensity of states rather than to international relations conditioned by economic needs. This point was, in his opinion, implicit in the very classical conceptions of imperialism, according to which colonialism was not always and everywhere undertaken to ensure the possibility for the investments of national capitals, but also to create sectors of the world economy protected by the state in which national capitals could operate "freely". The end of colonialism has led to a further territorial division and the number of states has risen dramatically. A world market has never existed in the true sense of the term and will continue not to exist because it is still mediated by the national state system.

Shaw held that the Leninist idea did not fully grasp the connection between the national state and imperialism. Lenin would have needed to combine his two studies, *Imperialism* and *State and Revolution*, in a single volume to do this. Certainly, Lenin had closely studied the links between industrial capitalism and banks, but he did not link this financial oligarchy with state finance and militarism. He emphasized the role of capital exports without indicating, aside for a few brief passages, how war had distorted the process of capital accumulation as he considered the expansion of militarism simply as a means of accumulating wealth.

The state dimension and its links with war, which were overlooked in *Imperialism*, were restored by Lenin in *State and Revolution.* But here too, despite what is stated in the premise, the connections are dealt with only in a superficial way, thereby, according to Shaw, confirming the outdated Leninist division between economy and politics and his final belief that the imperialist war would have vastly accelerated and intensified the process of the transformation of private monopoly capitalism into state monopoly capitalism, heightening the oppression of the working classes which in the end would unmask and destroy the state bureaucratic-military machine (Shaw 1984, pp. 53–55).

Lenin, like Hilferding, had analysed the functioning of the investment banks in the German model and their control over enterprises in order to stress the monopolistic role of the financial institutes of the time, calling this "finance capital", that is "the bank capital of a few very big monopolist banks, merged with the capital of the monopolist associations of industrialists". This merger came about because the banks are forced to sink an increasing share of their funds in industry. Thus to an ever greater degree the banker is being transformed into an industrial capitalist (Lenin 1916). Lenin summarized the essential characteristics of imperialism in 5 points:

> (1) The concentration of production and capital has developed to such a high state that it has created monopolies which play a decisive role in economic life; (2) the merging of bank capital with industrial capital, and the creation, on the basis of this "finance capital", of the financial oligarchy; (3) the export of capital as distinguished from the export of commodities acquires exceptional importance; (4) the formation of international monopolist capitalist associations which share the world among themselves, and (5) the territorial division of the whole world among the biggest capitalist powers is completed. (Lenin [1916] 1975, p. 83)

According to Lenin, the fundamental point was the relationship between imperialism and the monopolistic character assumed by capitalism at a certain stage of its development. Imperialism was nothing other than the external manifestation of the monopolistic structure, "imperialism is the monopoly stage of capitalism". That is the difference compared with Marx and Rosa Luxemburg, whose analysis of imperialism was undertaken within a theoretical framework of a competitive kind, even if Marx had shown how competition led to the concentration of production and this, at a certain stage of development, led to monopoly.

Another determining element in Lenin's analysis, as stated above, was that of the hegemony of finance capital:

> Imperialism is capitalism at that stage of development at which the dominance of monopolies and finance capital is established; in which the

> export of capital has acquired pronounced importance; in which the division of the world among the international trusts has begun, in which the division of all territories of the globe among the biggest capital powers has been completed. (Lenin [1916] 1975, p. 83)

While the old capitalism exported mainly commodities, the new capitalism under "the dominance of monopolies" exported mainly capital, which was accumulating to excess in the capitalist economies which had reached an "over mature" stage. This was the particular characteristic of imperialism.[5]

"Economic leagues", formed by industrial and banking groups, had substituted states in dictating international policy, thereby laying the foundations for the realization of a world capitalist economy which, unlike Kautsky, Lenin did not think would be able to establish an equilibrium. While Kautsky thought "ultraimperialism" saw monopolies cooperate in order to agree an optimal allocation of resources, Lenin believed that an agreement between monopolies was impossible because the relations of power between them was constantly changing. Antagonistic financial groups and trusts sought to extend their monopoly over territory with the support of their respective states. The export of capital was inevitable and implied the possession of zones of influence. The war grew out of violent conflict between national imperialisms over the division of the world, which could not be regulated pacifically.

In any case, imperialism was the inevitable product of capitalism, deriving either from problems of underconsumption, according to the theses of Rosa Luxemburg, or the need to avoid the fall of profits, according to the reading of Hilferding and Lenin, and led to war. Armed conflict was in any case present in class struggle in every form of society, except socialism. The ruling classes of individual countries needed to expand to maintain their domination and their opportunity to exploit the working class. Therefore, only a society without classes could finally eliminate war.

The Marxist theory of imperialism in its various interpretations provided an analysis of the causes that led to war based on economic factors connected with the capitalist mode of production, but at the same time, it assumed a prophetic vision of the end of war following the fall of capitalism itself.

This last aspect was generally abandoned by western neo-Marxists who have taken stock of the capacity of capitalism to adapt to changing economic circumstances. Samir Amin (1971) focused attention on capital accumulation on a worldwide scale and on the transformations that had taken place in the underdeveloped peripheral economies in relationship with the changes that took place in the centre. Charles-Albert Michalet (1976) analysed instead the characteristics of the contemporary globalized economy, where there had been a transition from international exchanges to the internationalization of production, thus realizing a new global form of imperialism.

Some economists with a liberal background brought their criticisms to bear on the theory of imperialism, disputing the basic hypothesis that real wages were destined to remain at a lower level. The capitalist system had in fact shown that it can raise the standard of living of the masses without destroying itself, so avoiding the proletarian revolution. Consequently, the entire theory would be discredited.

More generally, several economists made comments on the economic causes of war in the interwar period. In 1930, Ralph G. Hawtrey, an economist and friend of Keynes, held that every conflict was one of power and this depended on resources. Consequently, it was futile to seek to distinguish between the economic causes of the war and the other causes, because, in fact, the principle cause of war was war itself (Hawtrey [1930] 1952, p. 120).

On the eve of the Second World War, Lionel Robbins sought to define the role played by economic motives and particular economic institutions in causing wars in the modern period. He examined the Marxist thought on the matter, recalling that the Marxists had referred only to the historical period of capitalism. He showed that Rosa Luxemburg's explanation was wrong, while the Leninist explanation, even if it was valid in certain cases (as in the Second Anglo-Boer War) did not assume the character of a general explanation.

He considered that war was a complex phenomenon that did not derive exclusively from one cause. Economic motivations did underlie conflicts, but Robbins thought they were neither alone nor generally primary and that there was no interest, of a general character, in possible economic gains deriving from conflicts. In the long term, the interest "of the majority of the human race" did not lie in conflict. Wars were instead unleashed by the prospect of short-term advantages deriving from aggression.

At the end of the day, according to Robbins, war had essentially political roots, as it was the consequence of the existence of independent national states and their anarchic behaviour. It was not capitalism, as Marxists believed, "but the anarchic political organization of the world is the root disease of our civilization" (Robbins 1939, p. 94).

This statement showed that Robbins had largely fallen in line with the interpretation of the realist school that included among its more important exponents Raymond Aron and Hans Morgenthau.

Robbins attributed the theses of Hobson to the Marxists and held that the latter were right "in suspecting that in the causation of major wars there was sinister interest operating somewhere. But they mistook the nature of the interest". It was not the capitalists' greed that caused conflicts, but the interest of certain groups (that could be capitalists or workers or even formed by a temporary alliance of both) that gave rise "to the practices of restrictionism which are indirectly responsible for war" (Robbins 1939, p. 94).

Mises, in 1944, reflected on the war and the situation in Germany and state management in order to overturn the Marxist position on war, holding that "a lasting peace is possible only under a perfect capitalism", which nevertheless up till then had never existed anywhere or had even be attempted. In a free market world, where state activity was limited to the protection of life, of health and citizens' property, international conflict would not arise, nor could there be any economic reason for war (Mises [1944] 1974, pp. 317–318) The opposite was "Etatism", either in the form of economic intervention in the capitalist regime, or as socialism, which "inevitably must lead to conflict, war and totalitarian oppression of large populations" (Mises [1944] 1974, p. 106). Mises considered the peace plans outlined in the years he was writing and observed no will on the part of single states to abandon protectionism, seeing that the parties in power supported planning and public intervention in the economy instead. "They cannot demolish the trade barriers erected by their own countries. Thus the incentive for war and conquest will not disappear. Every nation will have to be ready to repel aggression. War preparedness will be the only means of avoiding war. The old saying *Si vis pacem para bellum* will be true again" (Mises [1944] 1974, p. 286).[6]

Apart from the economists, Hannah Arendt analysed the factors contributing to the birth totalitarianism and took into consideration the imperialist tendencies of a part of the European bourgeoisie at the end of the nineteenth century. Like Lenin, Arendt observed the presence of an excess of European capital that sought an overseas outlet. However, after paying attention above all to the political aspect of imperialism, she concluded by overturning the Leninist interpretation.

According to Arendt, imperialism was not the last stage of capitalism, but rather "the first stage in political rule of the bourgeoisie. [...] In the era of imperialism, businessmen became politicians and were acclaimed as statesmen, while statesmen were taken seriously only if they talked the language of successful businessmen" (Arendt 1951, p. 138). For Arendt, imperialism was the product of excess, of money and people too. The excess of money derived from the enormous increase in wealth obtained by capitalist-industrial production within a social system based on the poor distribution of income, agreeing here with Luxemburg. This wealth, which appeared to be in excess because it was badly distributed, in turn caused an excess in saving that could not be invested usefully within the economy of the national state, its organization and dimension at the end of the nineteenth century being no longer adequate for the needs of the capitalist economy. This made it essential for the bourgeoisie to assume political power and aim at imperialist expansion to be able to absorb the excess savings and people, rescuing capitalism from collapse.

Arendt meant by excess of people, those workers remaining unemployed during periods of crisis inevitably following phases of economic expansion. Then the exporters of capital asked for political protection too (Arendt 1951, pp. 147–149).

According to Arendt, the bourgeoisie had historically addressed the conquest of wealth, setting aside for a long time the assumption of political power, leaving this in others' hands on condition that their property would be protected. The problem arose when the excess of production deriving from the process of industrialization could no longer find sufficient outlets within the dimensions of the state. Then "when the accumulation of capital had reached its natural, national limits the bourgeoisie understood that only with an 'expansion is everything' ideology, and only with a corresponding power-accumulating process would it be possible to set the old motor into motion again". In other words, the bourgeoisie felt the need to start a process "of accumulation of power" to defend the accumulation of capital, determining the progressive ideology, characteristic of the late nineteenth century, and starting down the road to imperialism.

On his own part, "the imperialist-minded businessmen, whom the stars annoyed because he could not annex them, realized that power organized for its own sake would beget more power".[7]

What made the process irresistible was "the realization that power accumulation was the only guarantee for the stability of so-called economic laws" (Arendt 1951, pp. 143–144).

"The owners of superfluous capitals were the first section of the class [bourgeoisie] to want profits without fulfilling some real social function – even if it was the function of an exploiting producer" (Arendt 1951, p. 150). Colonial expansion offered this group a way out and saved the bourgeoisie as a whole from the consequences of the bad distribution of wealth.

The holders of excess wealth provided work for the workforce, again in excess, that emigrated from the four corners of the earth. In Africa, imperialism, "the product of superfluous money and superfluous men, began its startling career by producing the most superfluous and unreal goods": gold and diamonds (Arendt 1951, p. 151).

In the case of Great Britain, the colonies were not just

> a vast system of outdoor relief for the upper classes' as James Mill could still describe them; they were to became the very backbone of British nationalism; which discovered in the domination of distant countries and the rule over strange peoples the only way to serve British and nothing but British, interest. (Arendt 1951, p. 154)

The Department of Social Sciences of the University of Chicago promoted a project of studies in the mid-1920s aimed at investigating the causes of war with a multidisciplinary, non-theoretical approach. The research considered, among other things, the influence of economic activity on international conflicts and in particular considered capital investment abroad, which many authors judged to be a frequent source of international tension and controversy and believed that it tended to create an expansionist attitude.

Capitalists from industrial countries sought opportunities for investment in underdeveloped areas and made their governments offer diplomatic, naval and military assistance, often creating hostility with local authorities and even wars with other imperialist states.

Eugene Staley, a Professor at Stanford University and a future UNRRA project assistant, looked at the relations between private capital investment and international policies, in particular the way they generated tension between states. He analysed many concrete cases, both contemporary and historical, and inquired which of these services could offer private investment in the national interest abroad and which services, on the other hand, the state could offer its citizens who invested abroad. However, above all else, he sought to understand which relationship existed between investment and expansionism. In the end, he sought measures suited to resolve the political problems posed by private capital investment in foreign countries.

As regards aid, Staley concluded that in the concrete cases he had analysed, capital investments abroad had often been more useful to the United States Navy when it had been used to help in the protection of investors in foreign countries. He observed in some cases, it was the governments who asked capitalists to invest in underdeveloped zones in order to furnish a pretext to carry through intervention policy objectives (Staley [1935] 1967, pp. vi–vii).[8] But above all, he observed that expansionism was not in reality connected to an excess of domestic capital that needed to be invested abroad and, consequently, the pressure of excess capital could not be considered as a necessary cause of colonialism. It was more the case that capital was a means to realize it. The most important cases of expansionism without an excess of capital regarded Italy and even more so Russia. Italy and Russia started on an expansionist policy in the late nineteenth century and the first decade of the twentieth century, even though both countries had a shortage of capital. In both cases, expansionism was rooted in political ambitions and was dictated by the will for prestige, by military pressure and by "dynastic megalomania". Private capitalists obtained state incentives to intervene financially in the colonies and the state itself had already invested capital there, using foreign loans. For example, Russian capital invested in the colonies was in fact French capital.

In other cases, there were different reasons for expansionism. To a greater or lesser extent, it could effectively help dispose of excess capital seeking good investment opportunities. Jules Ferry in France used this argument to support the need for colonialism, while French bankers, according to Staley, did not exercise pressure in this direction. The reasons for French colonialism have to be sought in different factors: in the conviction that there was a need to find new markets for domestic production at a moment of generalized protectionism, in the will for prestige and more moral than economic compensation after the serious defeat in 1870 and in the conviction that France should have a cultural mission in the world. Staley agreed with Hobson as regards Great Britain in holding that the interest of private capital employed in various parts of the world had contributed significantly to the revival of imperialism in the late nineteenth century. Bankers and ship operators in Germany put on pressure in favour of colonial conquests, which in the eyes of the population appeared to be justified on the basis of national pride and the need for prestige. In all these cases, the interests of private capital seeking an outlet had played a more or less important role, but not exclusively.

Staley's research revealed that investors abroad sought purely economic gain and by acting alone drew the great powers into serious conflicts. On the contrary, the protective intervention of the state, meaning diplomatic protection of citizens and their interests abroad, sometimes increased or extended conflicts instead of sedating them. This was because, by adding the question of prestige to private disputes, it made their resolution more difficult as it stimulated hostile propaganda and furnished pretexts for carrying out aggressive policies of penetration (Staley [1935] 1967, p. 452).

Arthur Salter, in his introduction to Staley's book (Salter [1935] 1967, pp. xii–xiii), reinforced this concept, showing that the author had demonstrated how the correct economic motivation for foreign investment of capital could be transformed into something very dangerous when it became associated at a political level with ideas of power and prestige.

Quincy Wright, the author of the preface to the same book, instead stressed how Staley, while investigating the influence exercised by private investment abroad on international tension, came to illustrate a general condition in the contemporary world, in which the objectives and methods of national policy generally do not fall in line with a society that has become global in areas ranging from communications to trade.

Staley hoped for a further development in political, administrative and international legal control to solve this difference in the rate of change of economic evolution and the subsequent inevitable political adaptations, recalling that economic activities were those that required most attention in attempting to avoid war. He proposed the creation of world economic institutions to fulfil this aim, able to smooth out disagreement, and in particular suggested the constitution of a World Investment Commission and a World Commercial Court, a World Investment Bank, "International Corporations", that implied the creation of "international corporation law" and finally a world consular service. Staley was all the same conscious of the fact that these types of body could have only been able to function alongside a world political institution with powers of enforcement far superior to those exercised by the League of Nations.

William Langer, a historian by training and Professor at Harvard, but also Head of the American Research and Analysis Office for Strategic Services, later Vice Director of the Central Intelligence Agency, criticized the framework of Hobson and his "neo-Marxist" followers in the mid-1930s. He believed that the export of capital had only a weak link with territorial expansion. He took France as an example, observing that this country exported large amounts of capital throughout the world before the First World War, without this increasing expansionism. Just like Staley, Langer too recalled the declarations of Jules Ferry, but to stress in reality that French investments, even after the new territories had been annexed, continued to prefer Russia, Romania, Spain, Portugal, Egypt and the Ottoman Empire as destinations. Another example, that of Great Britain: in 1913, Britain invested more capital in the United States than in all of its colonies (Langer 1935, p. 103).

## Notes

1. Andersoncontested the Marxian idea that capitalism would have progressively reduced the importance of the nation state, thanks to its dynamic and universalist tendencies, to proceed in the direction of internationalism. If we look at the facts, the development of capitalism had instead reinforced nationalisms, according to Anderson, a point Marx did not examine deeply (Anderson 1976, p. 115). More recently, Hobsbawm (2011) instead rehabilitated the Marxian view of capitalism as a force which pushes towards the disaggregation of the nation state.
2. Kenneth Waltz (1959, p. 146), a leading light of "structural realism" (see paragraph VII.1), criticized the reading of Hobson, considering it to be excessively simplistic. "When he labels the foreign policy of capitalist states 'folly', Hobson is on strong ground, for there is ever a large admixture of folly in the acts of men and states. But in making imperialism the only folly of importance and in tying the imperialist folly entirely to the aims of a willful and a selfish minority, he has traded the materialism of Marx's dialectic for a materialism at once more naïve and less serviceable. This is, in short, single-cause explanation in one of its less impressive forms. The superficial virtue of the single-cause explanation is that it permit a simple, neat solution. Precisely so in this case".
3. According to Lionel Robbins (1939), this was not the observation that lay at the basis of Luxemburg's thought, who probably had got it from Rodbertus who, in 1858, when dealing with overproduction and crisis, had stressed the need for colonization as a response to the problems of underconsumption.
4. The article entitled *Imperialism* was first published in "Die Neue Zeit" XXXII, 1913–1914, pp. 908–922. [English translation in *Discovering Imperialism*, Chicago, 2012, pp. 753–772] As we shall read later, according to Hannah Arendt, late nineteenth-century imperialism was not the final stage of capitalism, but the first stage of the political role of the bourgeoisie.

5. For Lenin, colonialism would instead have had as its fundamental objective the seizure of raw materials by the more developed economies.
6. Mises, like Keynes, deplored the way the Versailles Peace Treaty had been pushed through which presaged a new war. "For centuries it had been the custom to conduct peace negotiation in accordance with the usages of gentlemen. The delegates of both parties, the victorious and the defeated, would meet as civilized people meet to conduct business. The victors neither humiliated nor insulted the vanquished; they treated them as gentlemen and equals. [...]. The Allied Powers broke this usage. They took delight in treating the German delegates with contempt and insults" (Mises [1944] 1974, pp. 286–287).
7. Reference is to a phrase by Cecil Rhodes, reproduced by Hannah Arendt (1951, p. 124): "the stars... these vast worlds which we can never reach. I would annex the planets if I could".
8. Here an episode comes to mind in the case of Italy (not mentioned by Staley) that was never completely cleared up: the purchase in 1869 of the Bay of Assab formally by the Rubattino shipping line in order to establish a coal bunker. In 1882, the bay was transferred to the Italian state and from there began its conquest of Massaua three years later.

## References


Amin, Samir. 1971. *L'accumulation à l'échelle mondiale: critique de la théorie du sous-développement.* Paris: Anthropos.

Anderson, Perry. 1976. *Consideration on Western Marxism.* London: New Left Books.

Arendt, Hanna. 1951. *The Origins of Totalitarianism.* New York: Hartcourt Brace.

Bukharin, Nikolai Ivanovich. [1918] 1929. *Imperialism and the World Economy.* London: International Publishers.

Comte, Auguste. 1830–1842. *Cours de philosophie positive.* 6 vols. Paris.

*Discovering Imperialism.* Series: *Historical Materialism.* Reprint 2012. Chicago: Haymarket Books.

Hawtrey, Ralph George. [1930] 1952. *Economic Aspects of Sovereignty.* London: Longmas, Green and Co.

Hilferding, Rudolf. [1910] 1981. *Finance Capital: A Study of the Latest Phase of Capitalist Development.* London: Routledge and Kegan Paul.

Hobsbawn, Eric John. 2011. *How to Change the World: Marx and Marxism.* London: Little, Brown.

Hobson, John Atkinson. 1900. *The War in South Africa.* London: Nisbet.

———. [1902] 1988. *Imperialism. A Study.* London: Unwin Hyman LTD.

Kautsky, Karl. 1907. *Socialism and Colonial Policy.* English translation in MIA (Marxists Internet Archive).

———. 1914. *Der Imperialismus.* "Die Neue Zeit" XXXII (II): 908–922.

Langer, William. 1935. A Critique of Imperialism. *Foreign Affairs* 15: 102–115.

Lenin, Vladimir Ilyich Ulyanov. [1916] 1975. *Imperialism, the Highest Stage of Capitalism: A Popular Outline.* Moscow: Progress Publishers.

———. 1917. *The State and Revolution.* Marxists Internet Archive.

Luxemburg, Rosa. [1913] 2015. *The Accumulation of Capital. The Complete Works of Rosa Luxemburg.* Peter Huds and Paul Le Blanc. London and New York: Verso.

Marx, Karl. 1859. *Contribution to a Critique of Political Economy.* Moscow: Progress Publishers.

Michalet, Charles-Albert. 1976. *Le capitalisme mondial.* Paris: Presses Universitaires de France.

Mises, Ludwig von. 1944. *Omnipotent Government. The Rise of the Total State and Total War.* New Haven: Yale University Press.

Robbins, Lionel Charles. 1939. *The Economic Causes of War.* London: Jonathan Cape.

Rodbertus, Johann Karl. [1858] 1969. *Overproduction and Crises.* New York: Kelley.

Salter, Arthur. [1935] 1967. *Introduction* to Eugene Staley, *War and Private Investors.* New York: Howard Ferting.

Shaw, Martin. 1984. *War, Imperialism and the State System: A Critique of Orthodox Marxism for the 1980s.* London: Macmillan.

Sismondi, Charles Léonard Simonde de. [1819] 1827. *Nouveau Principes d'économie politique, ou de la Richesse dans ses Rapports avec la Population.* Paris: Delaunay.

Staley, Eugene. [1935] 1967. *War and Private Investors.* New York: Howard Ferting.

Waltz, Kenneth N. 1959. *Man, the State and War. A Theoretical Analysis.* New York and London: Columbia University Press.

Wright, Quincy. 1935. *Preface* to Eugene Staley, *War and Private Investors.* New York: Howard Ferting.

# 5
# The Cost of War

Many economists in the twentieth century, following the two world wars and the upheavals caused by them, chose to focus on the analysis of the modifications brought by conflict to the economic system rather than on researching into the causes, both economic and not, that make wars break out, and evaluated the massive effects of state intervention in the economy in wartime, thus becoming accountants of war by calculating its direct and indirect costs and by seeking out the best way to finance it.

On the basis of such studies and following the extremely high costs of the Second World War, state accountancy in Europe reached a more precise level, aiming as well to establish with the greatest possible accuracy the quantity of entries that could be obtained through taxation.

James T. Shotwell, one of the promoters of the ILO, who aimed to see the inclusion of the Declaration of Human Rights in the Charter of the United Nations, coordinated wide-ranging international research (Shotwell 1921–1934) promoted in the first post-war period by the Carnegie Foundation for International Peace with the aim of attempting to measure by means of historical research the economic cost of war and the upset that it caused in the process of civilization (Shotwell 1927, p. XXI). Einaudi too took part in this research for Italy.



R. Allio, *War in Economic Theories over Time*,
Palgrave Studies in the History of Economic Thought,
https://doi.org/10.1007/978-3-030-39617-6_5

Many other contributions were published on the economy and war finance, including outstanding ones by Keynes, Einaudi and the lessons held by Lionel Robbins at the Marshall Foundation in Cambridge in 1947 (Robbins 1957), which we will come back to later, the book by Clark et al. (1918), and studies which instead drew up a balance sheet for colonial policy included those of J. H. Jones (1915), Grover Clark (1936b), and Emanuel Moresco (1937, 1939).

Michal Kalecki, a Polish Marxists economist, dealt with many aspects of the Second World War economy—inflation, price control, the purchasing power of wages and salaries, the reduction of small savings and the public budget—examining each problem in turn in articles published in the "Bulletin of the Institute of Statistics" (cfr. Kalecki *Studies in war economics* 1947).

Attilio Cabiati, an Italian laissez-faire economist, a friend and colleague of Einaudi's, considered the costs, principally the means for financing the wartime economy, after the First World War and the beginning of the second one. He went beyond just calculating the immediate effects of wartime taxation to consider also the long-term conditioning resulting from the choice of a certain type of financing rather than another one. He paid attention to the fact that the heavy taxation imposed during the war could in reality be "transferred" and the real contributor could in reality be different from the one hit formally speaking. His 1941 study also referred to Keynes's *How to Pay for the War* and he favoured his proposal, albeit showing that it was more effective the richer a country was and, therefore, could count on considerable savings.

Cabiati thought that war changed the international division of labour and hindered the functioning of the theorem of comparative costs. War and its "shadow" therefore did not permit reaching the highest level of general utility, destroyed wealth and favoured anti-economic production. Furthermore, the war made it impossible to achieve the goal itself of the economy, which is to obtain "the greatest advantage with the minimum effort", something, however, that for Cabiati did not just concern the economy, but "constitutes one of the natural and fundamental bases of worldwide development" because "all the physical laws in nature operate automatically following the principle of minimum effort" (Cabiati 1941, pp. 55–56).

In the 1930s, above all after Hitler began to make territorial claims, different economists researched into the costs and the problems connected with the possibility of a new outbreak of war, which inevitably would have been characterized as a total war. The works of Einzig (1934, 1939), Macfarlane (1937), Durbin (1939), and Clarke (1940) and the papers collected by G. D. H. Cole in 1938 were cases in point.

## Real Costs

Simon Kuznets stood out among the economists who sought to quantify the effects of war on the economy as the pioneer of historical evaluation of national income. During and immediately after the Second World War, Kuznets developed a series of statistics in order to attempt to define the changes that were taking place in the American national product and in the use of resources available following the conflict. Obviously, he encountered considerable difficulty in obtaining data and constructing price indices, above all those of war industry products. He made clear the problems that appeared in defining exactly and in aggregating peacetime and wartime production in the Gross National Product, given the different ways of evaluating prices, seeing that different institutional mechanisms determined them. Kuznets constructed several alternative series during the war using nominal values to then propose different deflation indices for expenditure on the war. At the end of the war, he revised his estimates to produce a picture of the American economy in the 1940s that was generally less positive than the officially published version (Kuznets 1942, 1944, 1945, 1951, 1952, 1961).[1] Kuznets held a course of university lessons in the 1960s where he attempted to evaluate the overall costs of the Second World War, including human and economic losses, effects on technological change, prioritization of choices and the organization of global structures.

Kuznets adopted a "pacific" conception of GDP in his studies into long-term growth which included only final products for consumption or new capital formation. From this point of view, defence expenditure can only form part of the GDP insofar as the stock of military capital could have been reconverted for peacetime uses. He considered war and defence expenditure as technically an intermediate product from a social point of view as it does not lead to "consumer's pleasure directly", but is instead "a necessary expenditure to 'produce' an economy" (cfr. Edelstein 2000, p. 387).

The Gross Domestic Product rose during the war, if measured conventionally, but this did not lead to an increase in welfare, so to measure this, war expenditure had to be deducted from the gross domestic product.

When dealing with the strong post-war growth and economic miracles, Kuznets (1964) stressed the difficulties encountered when defining those aspects of the war which could have influenced the subsequent models of a peacetime economy and the extent of the impact of these effects, both because certain data was not available and because adequate methods of analysis had not been formulated. He added that it seemed practically impossible to give a weighting to wartime activities and experiences in terms comparable to those of peacetime activities, that is to say "to translate somehow the horrors of war into dollars and cents; and the very attempt to do so may seem inappropriate, since it is bound to disregard the very essence of the effect of war on human beings and societies" (Kuznets 1964, p. 70).

Kuznets sought especially to quantify three effects induced by the Second World War: the level of economic losses, changes that took place in technology, institutions and the scale of priorities and, lastly, changes in politics and global structures.

It was already quite a task to calculate exactly how many human lives were lost, but even more difficult to make calculations of an economic kind and, above all, it was not at all easy to establish what could be the importance and the value of technological innovations that could not be realized following the wartime effort and which form the opportunity cost of technology (Kuznets 1964, p. 80). It is here that Kuznets used a concept contrary to wartime spin-off, proposed by Clive Trebilcock a decade later while attributing to technological innovation realized in the armaments sector the capacity to be reconverted and to innovate technology in peacetime production.

A wide range of research was published in the wake of the Second World War aimed at stressing the damage and at quantifying the costs of individual conflicts. Often the work was carried out by non-economists who did not use deflationary indices for short-term conflicts. It is impossible here to take into consideration such a vast amount of literature, but I feel it is worth recalling a famous book published some years ago by Bilmes and Stiglitz (2008) in which the authors examined with economists' methods the direct and indirect costs of the war in Iraq, a war they strongly opposed, reaching the figure of three trillion dollars.

Firstly, the two authors stressed how many of the political and economic problems that emerged during and immediately after the war were already quite clear, foreseeable and foreseen by those who had opposed the war before its outbreak. The book opened with the statement that "Now it is clear that the American invasion of Iraq was a terrible mistake", to follow on later by stating that the failure was not due to a single error, but instead to the host of dozens of errors that built up over the years, such that if someone wanted to analyse the failures of the American government, the invasion of Iraq could provide an excellent case in point (Bilmes and Stiglitz 2008, pp. IX, XIX).

The book listed the dead, wounded, and invalids, the net worsening of the living conditions and health of the Iraqi population and refugees, the destruction of the country's infrastructure, the rise in internal violence and in international terrorism, finally coming to the figure of three trillion dollars in material costs for the United States, perhaps to be doubled if we had included the costs for the rest of the world. But this was not all: we had to take into account the increase in the American public sector debt too, the increase in oil prices and energy in general and therefore the impoverishment of the economy, and finally the need to provide medical and psychiatric care for veterans.

The money spent on this war could have been used to cut taxes on the middle classes that were going through a period of economic difficulties, on the American medical assistance system, which was notoriously insufficient, so it could have been improved, or perhaps the money could have been better spent in Afghanistan to get out of that mess instead of getting into another. All in all, there were many ways to use the resources invested in Iraq in making America safer, more prosperous and better prepared to face future threats. But instead we have a fresh war, because "war is big business" (Stiglitz and Bilmes 2008, pp. XI–XII). The invasion of Iraq offered excellent opportunities for private companies in the defence sector. The Department of State alone spent over four billion dollars on security guards in 2007. The employment of contractors, of which there were over 100,000 operating in Iraq, enhanced the opportunity to make illegal gains and increase corruption. In some cases, the brutality of contractors worsened the conflict, while their high pay reflected on that of the US military, further increasing war expenditure.

Just as back in the time of Keynes, Stiglitz and Linda Bilmes too had worked with the political system, being consultants for the Clinton administration for years, without, like Keynes, ever being listened to.

At the beginning of the invasion, the United States spent on armaments at least what the rest of the world put together did and Iraq had concluded a long war against Iran in which hundreds of thousands of Iraqis had died. The Gulf War followed with losses estimated between 75,000 and 105,000 soldiers. Then there were ten years of sanctions. The military victory was, of course, not in question, but the declared objective of the United States was not the invasion, but rather that of freeing the country from dictatorship. The local population, however, got to know the American military as a force of occupation, not of liberation.

The American government financed the war above all through public debt, which was already enormous, thereby transferring the burden of repayment to future generations too. Taxes were not raised and, in fact, those falling on the rich fell. The cost of the war in Iraq, the costliest American conflict since the Second World War, was initially estimated to be 50 billion dollars. The three trillion-dollar figure calculated by Stiglitz and Bilmes in 2008, however, did not include defence expenditure, which was kept separate.

In reply to those who considered that the war had also presented economic gains, Stiglitz and Bilmes recalled that the price of oil at the beginning of the war stood at 25 dollars a barrel, rising to 90 dollars by 2007, that public demand certainly rose, but if the money uselessly wasted on armaments had been spent on infrastructure, plant, research, health and education, it would have improved growth and increased output. In fact, not by chance the American economy worsened during the war and produced below potential capacity. The economic damage to the community, in the authors' "realistically moderate" view, stood at over a thousand billion dollars (Stiglitz and Bilmes 2008, p. 115). America, by occupying Iraq, had instead effectively succeeded in enriching the Defence contractors[2] and the oil companies, first of all Exxon-Mobil, whose profits and share price took off, making them the real winners of the war. To conclude their research, Stiglitz and Bilmes expressed the hope that, when facing up to the desolate scenario, the American government could at least reflect on its errors and correct the political set up, proposing reforms for the future going in that direction.

## Opportunity Costs

A great debate arose and many calculations were made, mainly by non-economists, during the years of the Vietnam War concerning opportunity costs, the costs of the war measured in terms of opportunity, the usefulness that could have been obtained if the money spent on armaments had instead been used in works of peace. One of the main areas of economic damage of the war was identified just in its diversion of capital, labour and scientific research away from civil production towards the realization of unproductive goods and services. This concept was later also applied to the analysis of previous wars and conflicts in general.[3]

The sectors where the capital spent on armaments could have been invested were identified as schools, health, housing construction and welfare in general, which could have allowed for an increase in overall production. A summary along these lines can be found in many of the contributions on the war economy edited by Seymour Melman in a volume published in 1971.[4] The contribution by Bruce M. Russett, in fact, looked at opportunity costs and calculated to what extent American military expenditure in the post-war period led to budget cuts in the education and health and welfare sectors, highlighting just how many civil activities were cancelled due to military expenditure, concluding that, even if at the time of writing it was still impossible to evaluate the total costs of the Vietnam War, it was already clear that the impossibility of making adequate investment in civil sectors would have made Americans poorer, more ignorant and less healthy than in the past for many years to come (Russet 1971, p. 159). Similar conclusions were drawn in the same volume by Terence McCarthy when he stressed that the "invisible" costs of the Vietnam War could be just as serious as material ones, listing among the "invisible" costs, the instability of the dollar, the weakening of the political, diplomatic, military and even the economic position of the United States on a world level, with the parallel rise in Soviet influence (McCarthy 1971, pp. 161–162).

Louis Lundborg, President of the Board of the Bank of America Corporation and the Bank of America N.T. & S.A., opposed the widespread belief that the war was the motor of rapid growth and maintained that, on the contrary, increasing military research could have reduced the American rate of growth, considering the fact that the transfer of scientific and technological progress from the military to the civil sector had also become increasingly difficult. So if war did not advance scientific progress, it was of no use to the business sector either. Facts show that the Vietnam War seriously distorted the American economy, increasing inflationary pressure and draining resources that were desperately needed in dealing with internal problems, and lastly reducing the general increase in the rate of profit (Lundborg 1971, pp. 176–177).

American expenditure on defence and the space programme stood at 520 billion dollars between 1957 and 1966, two and a half times the amount spent on primary and secondary education, both public and private, or a quarter more than outlay on plant and machinery in the entire civil production system. But, as Arthur Burns added in the same volume, bombs and missiles did not increase productive capacity, while civil plant would have increased future production. The real cost of the defence sector was therefore not only that of the reduction of the production of goods and services, but also that of the lack in growth that would have taken place with productive investment in human capital and civil industry. Furthermore, federal financing of military research channelled to university institutes gave them a strong stimulus, but the finance was concentrated in certain sectors to the detriment of others and made research less independent. More widely speaking, investment in the defence sector complicated financial problems, making the economy increasingly dependent on the state, thereby profoundly changing the character of society itself, which became less free and more controlled.

Lastly, despite the exorbitant expenditure on armaments and the deployment of nuclear arms, the frustration of the Cold War and the brutality of the Vietnam War made Americans more insecure than their parents and grandparents ever had been in the past (Burns 1971, pp. 111–121).

Michael Edelstein calculated the direct and indirect costs of American wars in the first seventy years of the twentieth century: the First and Second World Wars, the Korean and Vietnam Wars and the Cold War too, and analysed the means for financing them: taxes, loans and printing money. He stressed opportunity costs, checking to see how private consumption of goods and services was wantonly sacrificed to finance all those wars and drew the conclusion that, even if the GDP measured conventionally clearly rose during wartime, this in no way led to a net increase in welfare (Edelstein 2000, p. 387).

Robert Higgs, in his article in the "Journal of Economic History" in 1992, forcefully rebuffed the then quite widespread interpretation that America went through a period of prosperity during the Second World War with a consumer and investment boom that definitely put an end to the effects of the 1929 crisis. According to Higgs, it was, on the contrary, the end to the war, the change in economic perspectives and the possibility to perform industrial reconversion that led to a return to well-being at the end of the war.

Higgs selected above all a quotation from Mises in order to deal with the improvements war could have caused. Mises considered wartime prosperity as the same as the prosperity caused by an earthquake or a plague (Mises [1919] 2006, p. 127). When referring to the First World War, Mises had in fact stressed that the apparent prosperity of the wartime period, connected on the one hand to the reduction in unemployment and unsold stock, and on the other to the increase in company profits, was in reality due to inflation whose effects were underestimated, so obscuring the real destruction of capital the war was leading to. Clearly, the destruction of goods and values could not lead to an increase in wealth. Mises listed some of the losses to the world economy caused by the war:

- The damage caused by excluding countries at war from the world economy
- The destruction of goods by military actions and the irreparable wear to military equipment
- The decrease of productive labour following the call up of civilians
- The reduction of the labour force due to war victims and the reduced working capacity of many ex-combatants

These losses were hardly ever compensated for by improvements to the process of production that could arise at the end of the war, nor by the application of innovative technology tried out in the military sector in wartime to the civil sector. Nevertheless, while dealing with the disadvantages arising for the general economy, Mises did not exclude the possibility that war, even in the contemporary period, could offer advantages to the victorious states when they were able to impose war reparations on the defeated nations for damages higher than the real damage caused. Militarism used this argument, even if there was no certainty a priori that war pays.

War is useful for armament manufacturers, which, nevertheless, according to Mises, did not influence the decision to declare war, while the greater profits for arms manufacturers did not compensate for losses in other sectors at the level of the economy as a whole (Mises [1919] 2006, pp. 125–130].

Higgs took as his starting point the arguments employed by Mises and believes that the evaluation of a positive effect of war on citizens' well-being could only derive from statistics obtained in a way that was neither accurate nor appropriate. A more careful interpretation of the data would show that within the immense arsenal produced by the Second World War, the living standards of consumers worsened dramatically. Real prosperity returned after the war. Certainly unemployment fell during the war, but only following call up. In those years, people worked longer in factories for a wage that allowed them to buy fewer consumer goods of a distinctly lower quality and at a higher price, while certain types of consumer goods were unavailable. The costs of transactions had risen, taxes too, and transport was scarce. How could anyone talk about "wartime prosperity" under such "conditions"? This did not take into account that greater state controls over the economy hindered the free operation of market forces. What Higgs failed to mention is the fact that the theories of wartime prosperity were spread about in the United States, a country where the war was not fought, where, on the other hand, production rose due to the increased demand on the part of other belligerent nations. Any idea of prosperity could not have gained hold in the devastated zones of Europe and Japan.

When dealing with opportunity costs, Higgs referred to Kuznets, who thought that wartime military successes and the maintenance of the social structure were equally important for individual welfare, if not more so. However, Kuznets himself had sustained that it made little sense to speak about the protection of life as an individual economic service. This was, however, an essential precondition for the offer of civilian services and welfare. Instead, many authors had considered military expenditure as a substitute for welfare, in the long-term too, even when war was not being waged, as was the case during the Cold War.

William Nordhaus and James Tobin[5] adopted the same position as Higgs, even if only in part, when they stated that military expenditure could not be treated in the same way as welfare, but it could not be considered as a waste either, being instead "an unfortunate necessity" (Nordhaus and Tobin 1972).

Even if the editor of "The Economist", Francis W, Hirst, did not use the concept of "opportunity costs", as early as 1915 he stated that it was the plain truth to say that war impoverished people and nations. Both in the past and again today, it has been veiled by "artistic" representations, fed by special interest that sought to depict war as a gold mine. Hirst thus rebutted the three principal statements used at the time by others to sustain that war generates well-being. The first is that war produced wealth with the expansion of the circulation of money, the second is that the improvement was caused by demand increasing to replace goods destroyed during the war, while the third and the most widely used considered the advantages in the fall in unemployment. In reality, as Hirst recalled, the increase in circulation caused inflation and not everything that was destroyed would be rebuilt, either because there was no economic benefit in doing so, or because the owners were dead or did not have the financial means for reconstruction. Moreover, war did not increase employment, but was limited to moving workers from productive to destructive sectors. He found that the idea that high wartime tax rates were useful as they forced people to work more than they would have had to otherwise to be able to pay the taxes was a total idiocy.

Melman too rebuffed those who held that war brought prosperity, a conclusion reached by many Americans when they saw that the Second World War (fought elsewhere) had brought the effects of the Great Depression following the 1929 crisis to an end. Increased public demand for armaments had in fact led to full employment of productive factors, industrial potential was fully used and unemployment was drastically cut. But, as Melman stressed, public opinion did not see that the wartime economy had raised the public debt, favouring only certain companies and "bureaucratic capitalism", heightened the power of state intervention and tied it into militarism, making both the economy and society less free.

Moreover, economic support for military demand did not stop even in the years after the conflict, following a rapid rearmament due to the Cold War and the outbreak of the Korean War. This period saw between 7 and 10% of the American GDP spent on military purposes (Melman 1974, p. 15). At the end of the 1970s, America had more armaments than required to eliminate humanity totally and Melman noted with bitter irony that even military power must in the end reach its limits, if for no other reason than that "people cannot be killed more than once" (Melman 1974, pp. 162–163), just as a city or a state cannot be destroyed another time over.

On the other hand, no one has taken into account the fact that, in economic terms, the production of armaments was unproductive because it did not contribute to the improvement of living standards through the increase in consumer goods and services and did not even serve to increase future production (Melman 1974, pp. 24, 62). Melman complained in particular about the fact that this characteristic of military expenditure was never mentioned in university economics textbooks.

Paul Kennedy, while analysing the years of the Cold War and without explicitly referring to opportunity costs, recalled that when the Soviet Union and the United States spent hundreds of billions of dollars every year on armaments, some political observers began to ask if both powers had not already become "states of national security" concentrating obsessively on the argument of defence, while other researchers, who considered rather the economic aspect of the phenomenon, were worried about the massive diversion of capital, research and development, scientists, engineers and technicians from private sectors to the arms race and had begun to fear for the effects of this choice on long-term national competitiveness (Kennedy 1993, p. 167).[6]

## Paying for the War

The theorist of military thought, Jacques-Antoine Hippolyte Conte de Guibert, had already back in the 1770s begun to study the economic management of war and the problems of subsistence, supplies and transport during conflict.[7] He also paid attention to the relations that could arise between economic interests and strategic decisions.

Guibert was a case quite the opposite of those we are examining here as he was not an economist who considered the problems posed by war, but a military strategist who took account of the new economic conditions that cropped up during wars and the relationships that linked the economy and war.

Guibert (1772) held that war could only be understood if it is seen within a vast totality of phenomena and thought that the economy was an element in it. His treatises on military technique considered that knowledge of strategists of the economic situation of the country in defining decisions on the war as fundamental. The science of subsistence had to be tied in with the science of strategy. In fact, even if military strategy dictates the behaviour to be adopted, for each war account has to be taken of economic conditions and the availability of finance for the choices made.

The costs of war that the economy had to bear include expenditure to arm and maintain the troops and the costs of organizing the system of war and of guaranteeing its financing. If the economy was able to furnish the needs required to conduct the war, the conduct of the war would interact in turn with the behaviour of the economy.

In more recent times, the war economy, with its costs and problems and massive public intervention, has been examined by various economists including, apart from Hirst, mentioned above, Pigou, Keynes and Einaudi.

The traditional ways for financing war are all in all not so many: either the state increased taxation, thereby reducing domestic demand, or it ran up debts, placing the burden of the costs of war on the shoulders of future generations, or its printed money, causing inflation. However, most likely it did all three at once. During the war, the state could even reach the point of confiscating citizens' possessions or making them do forced labour.

Classical economists, starting with Smith, Ricardo and Say, generally held that it was preferable to finance war with income taxes rather than public debt, a move which lengthened the time taken to repay interest. Better still, of course, if the taxation fell on those who gained from the war. Nevertheless, they were all conscious of the fact that taxation alone was insufficient to face the costs of a long war.[8] This remained true even more so for extremely expensive modern wars. Charles Davenant, an English mercantilist, had already warned from the end of the seventeenth and the beginning of the eighteenth century (1698, 1701, 1704), that financing a "long" war required means other than taxation and on the matter expressed his strong opposition to long-term debt because its high interest rates would have been a burden on all the productive activities of the country for many years to come.

Hirst (1915) stressed how the level of public debt was necessarily limited by the extent of national savings and that taxation could not exceed the tax-paying capacity of citizens when he examined the ways of financing war employed by European states at the beginning of the twentieth century and how they affected the economy. On the other hand, printing money helped the state in the payment of its suppliers, but caused currency instability and devaluation. Hirst also considered another form of intervention for dealing with war expenditure which he called "confiscation" of labour or property, which he judged to be "anti-economic".

As regards labour, Hirst meant by "confiscation" the ways by which the wartime state could force workers and employees to work longer hours and force into work women and children and those already pensioned off. But he thought that "forced" labour was neither economic nor efficient and, in fact, not only this measure, but also forced conscription would be abandoned sooner or later for its anti-economic characteristics. As regards confiscation of property, Hirst noted that, in the times he lived, it was a real risk, because the feeling of the working classes during the war went in that direction. Consequently, the war posed a real danger "for property" and the War Minister could indeed be tempted or pressed by public opinion to push up taxation of great wealth, almost to seem like a confiscation. Hirst drew on examples of this type of action from fiscal measures taken in Japan and Russia and the decision made by Great Britain during the Boer War too.

Expenditure on security too must be kept under control and never become excessive. In 1911, Hirst calculated that the provision of funds to such an end made by the British government was altogether sufficient to maintain the Empire safe from invasion and, therefore, any further expenditure, which could be seen as potentially provocative, was to be avoided and even that all possible efforts were to be made to reduce the level of armament. Hirst in fact tended to view all expenditure on arms, ships and means of destruction as waste, while, until permanent peace had been reached, all expenditure on peace was absolutely necessary.

While condemning the war debt as worse than false wartime prosperity, Hirst dedicated much time to the analysis of the public debt that Great Britain, France, Germany and the United States were running up in those years and to the study of the effects that this major indebtedness could have on the economy in general. He therefore took into consideration the effects on the civil population of war expenditure, wartime inflation and of the transfer of funds previously used for civil requirements to expenditure on armaments, and so concluded by recalling that there was a limit, if not to the destructive possibilities of war, then, as Melman was to note later, to the human suffering and economic misery a state could inflict on its people as a consequence of war (Hirst 1915, p. 150).

In 1916, the year after the publication of Hirst's book, Pigou drew a distinction between the monetary costs and the real costs of the war, meaning by real costs, not the costs minus inflation, as usually was the case, but instead the opportunity costs, being the costs of goods for civil use that the community had to renounce in order to produce armaments, goods and services needed for the war.

As for the choice of the way to finance the war between mainly taxes or loans, according to Pigou it was not correct to state that taxation necessarily entailed payment from current resources, while a loan implied the use of future resources for repayment and interest. The matter of if the burden of the war was to be borne in the present or the future depends on where taxpayers or lenders obtained the money to pay taxes or make the loan (savings, private loans) and on how they would have otherwise used these monetary resources (consumption, productive investments). In any case, interest on loans would have to be paid from future taxation, which will also fall on the shoulders of those who acquired the bonds, so progressive taxation seemed to be preferable to issuing wartime bonds.

Keynes and Einaudi, who both, but in different ways, participated actively in the political life and choices of their own countries, also studied the costs of war and the way to pay for them, making criticisms and offering suggestions to avoid further damage, if not further wars.

In Chapter 10 of the *General Theory*, dealing with the multiplier, Keynes made a short remark about war. Given that "unproductive" expenditure financed by loans to the state (in a situation in which productive factors were not fully employed) determined a greater than proportional increase in production and income, Keynes placed war alongside natural disasters in determining expenditure which could, in the long run, offer some advantage to the economy in general. He took this occasion as an opportunity to shoot a barb at neoclassical economists: "Pyramid-building, earthquakes, even wars may serve to increase wealth, if the education of our statesmen on the principles of the classical economics stands in the way of anything better" (Keynes 1936, p. 129). Immediately after, he entered into a polemic with the supporters of the gold standard with his well-known example of uselessly digging holes in the ground. Obviously it would have been "more sensible to build houses and the like", but doing so often ran into political and practical difficulties, so digging holes was better than doing nothing. The same could be said of war, which was "the only form of large-scale loan expenditure which statesmen have thought justifiable". The money would certainly have been better spent, and so a contribution would have been made to reaching full employment and so ending the depression (Keynes 1936, p. 130).

Keynes had, nevertheless, already begun to work on and worry about war before becoming a "Keynesian economist". In fact, he never shared the idea of Wilson that the First World War was the war that ended wars. Instead, he thought that European economic unification was the real way to pacify the old continent. At the war's end, he was invited to participate at the Versailles Conference as the representative of the British Treasury. He resigned when he clearly saw that the politicians continued to worry about problems of frontiers and sovereignty without being able to evaluate effectively the effects of what they were hastily deciding. He went to his Sussex home and in less than two months wrote an intense and bitter book that was destined to obtain a great echo and cause many polemics, *The Economic Consequences of Peace*. He dedicated the book to the new generation which still had no say and had not yet been able to express an opinion, since those who lived through the war were exhausted:

> We have been moved beyond endurance, and need rest. Never in the lifetime of men now living has the universal element in the soul of man burnt so dimly. For these reasons the true voice of the new generation has not yet spoken, and silent opinion is not yet formed. To the formation of the general opinion of the future I dedicate this book. (Keynes 1919, p. 279)

The book was a ruthless accusation of the incapacity of Wilson (and of his messianic conception of the tasks of the United States) and a rebuttal of the "sophisms" and "Jesuitical exegeses" of Clemenceau and Lloyd George, tied in turn to the old policy of claims and revenge, reparations and annexations. Keynes contrasted the logic of war with the logic of the economy and accused the tragic political stupidity and the material economic impossibility of imposing the extremely hard reparations imposed on Germany in order to destroy its economy. This decision showed the incapacity of politicians both to consider not just single nationalities, but the future of Europe and its true interests, and to evaluate what effects the application of war damages on Germany would have had on the economies of the winners' own countries, the damages being so heavy that they could only be paid in natural resources, such as raw materials, and finished products too.

Keynes believed that the Peace of Versailles would not hold and proposed that it should be revised, that inter-ally debts should be regulated, and an international loan should be negotiated in favour of those countries which had been devastated by the war to allow them to make a fresh start, and finally to recast economic relations between Central Europe and Russia. His appeal fell on deaf ears.

Keynes considered the theme of war again at the beginning of the Second World War with his famous pamphlet *How to Pay for the War*, published in 1940. Here he returned to the proposal he had made the year before in two articles in "The Times" on 14th and 15th November in favour of a programme of enforced saving to be applied in Great Britain so as to avoid seeing the costs of the new war falling on the shoulders almost exclusively of salaried workers following inflation, as had happened in the previous world war. Keynes's first consideration was that workers would certainly work more during the war and earn higher wages, but that these wages would be hard hit by inflation, so that in the end they would be able to consume less than before. He then listed the ways that could be used to deal with the economic emergencies of the war, selecting three remedies and two pseudo-remedies commonly used to obtain financial equilibrium. The first pseudo-remedy was rationing consumer goods, which however turned out to be ineffective when purchasing power was increasing. The second pseudo-remedy consisted in measures to be taken against war profiteers who exploited to their advantage the negative situation favouring price increases. But, as Keynes saw it, whoever thought that these pseudo-remedies could have really been effective did not see the real nature of the problem that was the fact that wartime aggregate demand rose faster than the supply of goods (Keynes

1940, p. 43), the exact opposite of a situation of depression where aggregate demand was insufficient to meet the offer.

Turning now to consider the real remedies, the first was to simply let prices rise seeing that some increase was inevitable during wars. Keynes recalled here that despite all the efforts made to stop it, living costs in Great Britain rose by 6.5% in the first month of the war. But if everything was left to this remedy, the rise in prices required to restore equilibrium would be intolerably high and socially unjust as it would hit hardest people on the lowest incomes.

The second remedy was taxation. But any serious contribution to resolving the economic problems caused by the war would have to include taxing the working classes, hitting the consumption of low-income groups. So both of these remedies, price increases and taxation, would therefore prevent workers from benefitting from the increase in purchasing power obtained during the war.

The third remedy, the only one that did not cause this disadvantage in the long run was that found in the proposal of "deferred consumption" (a term which had a much less negative impact than "forced saving") that was to be carried through by the obligatory deposit made by all English taxpayers of a portion of their income in the Post Office Savings Bank to be kept there until the cessation of hostilities. The state would pay a low rate of interest on the forced savings.

Keynes felt sure that the mobilization for the war would have led to pay rises, but this should not lead on to an increase in consumption which would have fed inflation, all the more so because the availability of goods for civil consumption would have decreased with the increase of war production. Inflation would have made workers pay too high a price. The special aspect of the remedy he proposed lay in the fact that it splits the monetary compensation earned by everyone into two parts, one part to be paid in cash that could be used for immediate consumption needs, the other part with the obligation of deferral of use until the wartime emergency ceased and productive resources began to increase again. Keynes was aware of the fact that deferred payment of part of the salary or other source of income could be unpleasant for the earner, above all the worker, but held that it was the only way to be able to keep real wages stable. The alternative was inflation. By limiting immediate consumption instead, an excessive rise in prices was stopped and the consequent loss in the value of salaries. The system also provided financial resources to the state and therefore avoided an excessive rise in taxation.

Keynes calculated the percentage that could be saved weekly by the various income bands and family sizes and so form the enforced deposit, exception being made for the lowest income groups and for situations of proven spending requirements. The deposits would be freed and made available to the savers, probably in the form of instalments, a short time after the war. This would have provided workers with available cash that could have allowed the relaunching of private demand at the time that public demand was falling. However, Keynes seemed to have underestimated the difficulties the state would face at the end of the war in finding the financial resources for reimbursing the enforced loans of private citizens in an economy that was likely to have been devastated.

If we could say that the Keynesian hypothesis had some possibility of success in Great Britain, where wages at the start of the war were rising and could in effect allow for some savings, it would turn out to be objectively impracticable in countries with much lower per capita incomes and workers' salaries, such as Italy. In this latter case, above all with the war dragging on and the scarcity of essential goods, there was no possibility of making savings from wages as they only allowed with difficulty for the purchasing of basic essentials for survival, which were rationed too.

Keynes's proposal gave rise to a great debate in Great Britain and beyond.[9] Many economists came out in favour, including some who were far from holding Keynes's positions, such as Hayek, who went as far as suggesting a circular approving the project.[10] Keynes thought Lionel Robbins too, even if he did not say so publicly, was in general agreement with his idea, as were many public administrators, the Governor of the Bank of England and the Canadian Foreign Minister. The only economist who came out against the plan was John Higgs who opposed it from the pages of "The Manchester Guardian".

Nevertheless, Keynes's proposal was not followed up because it did not get the backing of the Labour Party and the Trade Unions and was not appreciated the majority of English public opinion either.

In the autumn of 1939, he had sent the manuscript of the articles published in "The Times" to Clement Atlee, leader of the Labour Party, expecting to gain his approval. Instead, there was a roundly negative judgement (Skidelsky 2005, p. 590). The Party's official reply was published in a pamphlet called *How to Pay for the War* (a title Keynes himself adopted in 1940)[11] signed by Evan Durbin in the name of an "Informal Committee on the Finances of the Nation" (Durbin 1939, p. 17).

According to Durbin, low taxation and the lack of control over industry would necessarily lead to seeking major loans with the consequent greater inflation. It would be better to have low inflation kept under control along with rationing and other limitations on consumption. He started from the consideration that industrial mobilization would divert resources from civil consumption to war production, and so asked how the government would be able to control the use of these resources to cover the costs of armaments. He therefore proposed six guidelines for public intervention, nevertheless noting the counter-indications in each case: a rapid rise in taxation (which nevertheless would cause discouragement and inflation in the private sector)[12]; direct control over industry by the state (even if, perhaps, there was not enough time to set up an efficient system), expansion of the money supply (that could, however, cause an inflationary spiral); low interest rates on government bonds; control over the loans made by private banks and the obligation of banks to offer loans to the state (even if enforced loans are an unpopular expedient).

The two different proposals, Keynesian and Labour Party, led Gavin Kennedy (1983, p. 20) to say that while the liberal, Keynes, tried to save the income of workers, the social-democrat, Durbin, tried to share out the sacrifices.[13]

Keynes's piece was taken into consideration abroad by economists who were called on by the respective governments of the countries involved in the war to suggest the most effective ways to provide financial support for the war effort. The Italian case was reconstructed by Giovanni Pavanelli (1989–1990) in a long article that appeared in two issues, in which the reports made by some economists regarding war finance were republished. In 1942, the Istituto Nazionale di Finanza Fascista, itself founded in 1939, set up a Comitato Tecnico with the task of analysing the various ways of financing the war. After a rapid look at Keynes's hypothesis, which was held to be interesting, but impracticable, the Committee went on to examine the model developed by their German ally, called "capital circuit" or "monetary circuit". This model foresaw the supply of liquidity that, however, was not be used to increase demand in order to avoid inflation. Therefore, it was necessary at the same time to block prices and maintain a rigorous control over production, so that any monetary surplus was directed towards investment in state bonds, or could be taken as taxes. The difficulties met in applying this model to the Italian case in wartime were stressed by the members of the Committee in March 1942. They highlighted the fact that the system implied strong state control, on the one hand, and the will and possibility of citizens to underwrite the growing public debt on the other, this second hypothesis being difficult to realize in Italy, given the low wage rates and levels of consumption of the workers. A member of the Committee therefore proposed a "general mobilization of wealth" (Pavanelli 1989–1990, II, p. 62) to be carried out through the requisitioning of companies, or parts of them, by the state, because the war was costly and could not be paid for by using traditional methods or various expedients. So a debate began over the alternative proposals for special taxation of surplus profits and the increase in company capital, and lastly, setting a limit to dividends, as well.

Attilio Cabiati, in his 1941 study of war finance in Great Britain, noted the high savings rate of the country and the relatively low level of taxation compared with other states, above all taking into account of the greater wealth of the country. These favourable circumstances allowed Great Britain to avoid, at least for the time being, both inflation and major foreign loans and even made the application of the "noted and intelligent suggestion of Keynes" (Cabiati 1941, p. 122) unnecessary.

The longer and more noted contributions by Einaudi to the question of war concerned wartime taxation and the economic handling and social effects of the First World War in Italy. But Einaudi also expressed his position regarding war in many articles published in "Corriere della Sera" and "Riforma Sociale" and even more so in certain interesting, but largely ignored, essays written at the end of the First World War and the conclusion of the Second World War. Einaudi enquired in these articles as to how it would be possible to avoid further wars and noted problems still current today. He suggested the creation of a European economic federation, like Keynes, both in 1918 and in 1943, in anticipation of seeing the birth of "the United States of the World", but he also noted, with preoccupation, the existence among European populations of a deep cultural bond to their fatherland.

Einaudi did not believe that the two world wars, as with wars in general, were the result of economic causes. When dealing with those who held that the global conflicts found their origin in the struggle of large industry to conquer foreign markets, both to distribute surplus products and to gain access to raw materials useful to manufacture the surplus products. Einaudi recalled that war had historically existed since well before industrialization and, even more so, before surplus production and, in any case, was simply due to errors in calculations made by producers. A free market would correct these errors automatically through the reduction in price of the glut of products, while entrepreneurs would shift their investment to other sectors, without causing collective damage.

The search for external outlets by violent means could only lead to the saturation, following that of the domestic market, to that too of the "vital space" that had been conquered, and then the world market as well. In this last case, "if an ineluctable force thrusts great industry to exceed the capacity of consumption, against who would the entire globe of land and water go to war to find an outlet for the so-called excess?" (Einaudi [1943] 1953, pp. 26–27). Einaudi further recalled that the concept of vital space, seen as the remedy for excess production, had always found students of economics to be sceptical, because it presupposed that those spaces aimed at economic self-sufficiency too. Furthermore, the violence employed in the conquest of the territories destined to become outlets for surplus production, itself due to errors of calculation, would only with great difficulty lead to a lasting peace between the subjected peoples in their role of consumers.

Einaudi then went on to condemn another idea too that had become widespread before the war, that the economy should be subaltern to politics, that is private economic interests should necessarily be subject to the interest of the state, its spirit of power and domination. Not only was it untrue, but exactly the opposite took place in fact with private economic groups bending to their needs the policies of states, so that "economic victory" was no longer won by the best producers, "but those most able to gain influence over the government" (Einaudi [1943] 1953, p. 33).

These erroneous ideas had contributed to increasing the conflicts of the first half of the twentieth century. If there had been true freedom of trade, the better producers would have been able to make way domestically, and in the international field the relationships between nations would have been peaceful. While the exact opposite was:

> Vital space, the pretence of the dependence of the economy on politics means [...] domestic preponderance of financial groups of exploiters of the working masses and consumers, wishing to form monopolies and expecting privileges within a closed domestic market and needing to stir up jealousy between states in order to prevent the competition of others. Freedom in international trade means peace, while vital space and the dependence of the economy on politics means international tension and war. (Einaudi [1943] 1953, p. 61)

But Einaudi also uncovered the presence, in 1943, of a conflict between political-cultural and economic factors that was analogous to the one that re-exploded after the fall of the Berlin Wall. While the economy even then required free markets potentially at a global level, "wherever peoples have been free to manifest their will outside the bounds of state pressuring, we can see the anxiety to reconstruct small countries, to reaffirm their independence to bring back regions to life, the little homeland, the religion of the language and native customs" (Einaudi [1943] 1953, pp. 34–35). Thus contrasting desires and tensions arose between populations which yearned for spiritual and cultural autonomy and technology which tended to unify the world economically and national governments, which broke this unity with various types of artifice.

Einaudi therefore saw the two world wars as "the tragic manifestation of the historical necessity for the economic unification of the world" (Einaudi [1943] 1953, p 38). He was pessimistic too, because despite the huge loss of blood and material destruction, he realized that the aspirations of the little countries would last. The problem that Wilhelm II and then Hitler had posed with violence had not been resolved. Other wars appeared inevitable before achieving "the unification of the world under the heel of a hegemonic people" (Einaudi [1943] 1953, p. 39). It was vital to obtain the economic unification of the world by pacific means in order to avoid this enormous tragedy. The road to take to achieve this was not a new League of Nations, bereft as it was of any coercive powers, but rather an economic federation, starting with Europe. Einaudi therefore drew up a scheme of this federation which should take care of international exchange, transport and communications, internal migration, currency issues, literary rights, patents, weights and measures and health regulations. He offered an example too of the administrative structure, a common army and the means in which it would possible to act concretely. In fact, Einaudi's project preconceived what the European Union laboriously realized in later years His essays on the question were collected and republished in Italy by the Movimento Federalista Europeo in 1953.

His two volumes of considerations of the First World War are of a different content. The first, concerning the Italian tax system during the war was published in 1927, while the second on the *Condotta economica e gli effetti sociali della Guerra italiana* [The Economic Conduct and the Social effects of the Italian War] appeared in 1933. Einaudi wrote them for the research work conducted by Shotwell mentioned above. They were promoted by the Carnegie Foundation for International Peace. Einaudi was the President of the Italian Committee for publication. The first was highly technical work in which Einaudi checked on the changes that had taken place in the Italian taxation system following the war: new taxes, new tax bands, exemptions and tax revenue. The volume also analysed the effects of war finance on wealth, but not on family consumption. The conclusion is that "if during the war and due to the war, the distribution of the tax burden shifted", in Italy the only major modification derived from the application of special taxes on war profits and the surtax on wealth (Einaudi 1927, p. 484).

The 1933 essay, as Einaudi himself wrote in the preface, was not a "complete" history of the behaviour of the Italian state and society during hostilities, but rather "a contribution of notes dictated by a contemporary concerning the line adopted" (Einaudi 1933, p. XXVII). Einaudi calculated war expenditure taken from the organization of the state machine for the control of production and for intervention in various productive sectors, this activity going under the name of "wartime collectivism". He protested against the "fiscal massacre" and the delay in setting up the Commissione Consultiva for the revision of prices which allowed auxiliary companies to take advantage of the situation by profiteering on the backs of taxpayers up to and throughout 1916. He took account of the shady dealing and considered the difficult return to post-war normality too, with the very high level of social tension, hostile bank takeovers, the occupation of farmland and factories and the problem of inflation. He did not even mention the rise of fascism.

## Notes

1. In later years, other indices for deflation during wartime were constructed by Milton Friedman and Anna Schwartz in a long-term study into the US currency (1963) and also by Hugh Rockoff and Geoffrey Mills (1987).
2. Many publications concerning contractors have appeared in recent years written by politologists and jurists to accuse the renouncing by the state of its monopoly of violence in employing them. Many authors have written about the interweaving reasons, both economic and not, underlying Bush Junior's intervention in Iraq, to ask to what extent the question of oil was involved, what were the economic interests of Bush himself as well as Cheney and American oil producers, what was the desire to control the Middle East and what was the role of American imperialism. Cfr. Above all D. Harvey (2005) and the short summary made by Michael Mann in "Incoherent Empire", in the chapter on the attack on Iraq (pp. 206–251).
3. The hypothesis that underlies this discussion is that of full employment of productive factors (meaning a choice has to be made between butter and cannons), accepted both by the Austrian School (Mises and Hayek) and the neoclassical school, but rejected instead by Keynes in his *General Theory*.
4. Seymour Melman was a member of a circle of critical radical intellectuals, along with Noam Chomsky and John Kenneth Galbraith, who promoted disarmament and economic reconversion in the 1960s and 1970s.

5. Both US economists: Nordhaus specialized in economic problems connected with climate change; Tobin won the Nobel Prize for Economics in 1981.
6. On the same argument, see also Adams (1941), De Grasse (1983), and Cohen and Wilson (1990).
7. Christian Schmidt (1991) in particular took into consideration the importance of Guibert in the role of economics in military strategy.
8. In his study of the economic conduct and the effects of the First World War (1933, pp. 40–41), Einaudi held instead that with an equalized tax system and by accepting a decade of high, but bearable, taxation, "Italians would have been able to conduct the war without remaining debts to pay, without monetary inflation, without price rises or falls, apart from ones inevitably following any transition from one productive goal to another and without serious instability for positions previously already gained".
9. Cfr. Keynes Collected Writings 1978, Vol. XXII, pp. 40–155, Skidelsky 2005, pp. 587–592.
10. This is probably due to the fact that on this occasion Keynes was thinking in terms of the full employment of productive factors in the same way as the Austrian and neoclassical schools of thought.
11. The same title had already been used at least twice before: in 1916 for a book edited by Sidney Webb and published by the Fabian Society and again in 1918 by the American financier Arthur Stilwell, who was a founder, among other enterprises, of the Kansas City Southern Railway.
12. Durbin holds that, at least in theory, it would be possible to finance the entire cost of war just through correctly imposed taxation, thus holding down inflation and not spreading costs over time. Such a method would, however, be extremely unpopular (Durbin 1939, p. 36). Einaudi had already expressed the same idea in 1933 (see note 8 above).
13. The Marxist economist Michal Kalecki instead asked himself "who pays for the war?" and challenged the method normally used to find an answer to the question, that is the calculation of the direct taxes paid by the different social classes, without taking into account the different weight of indirect taxes borne by the less well-off class and the rise in the prices of staple goods and transport, albeit admitting that this type of calculation was objectively difficult to perform (Kalecki 1947, pp. 221–222).

## References

Adams, Gordon. 1941. *The Politics of Defense Contracting: The Iron Triangle.* New Brunswick: Transaction Books.

Burns, Arthur. 1971. "The Defense Sector: An Evaluation of Its Economic and Social Impact." In *The War Economy of the United States*, edited by Melman Seymour, 111–121. New York: St. Martin's Press.

Cabiati, Attilio. 1941. *Il Finanziamento di una Grande Guerra.* Torino: Einaudi.

Clark, Grover. 1936a. *A Place in the Sun.* New York: Macmillan.

———. 1936b. *The Balance Sheets of Imperialism.* New York: Columbia University Press.

Clark, John Maurice, Walton Hamilton, and Harold Glenn Moulton. 1918. *Reading in the Economics of War.* Chicago: The University of Chicago Press.

Clarke, Richard William Barnes. 1940. *The Economic Effort of War.* London: Allen & Unwin.

Cohen, Richard, and Peter A. Wilson. 1990. *Superpower in Economic Decline: U.S. Strategy for the Transcentury Era.* New York and London: Taylor & Francis.

Cole, George Douglas Howard. 1938. *The Economic Consequences of War Preparation.* London: Persons & Periods.

Davenant, Charles. 1698. *Discourses on the Public Revenues and on the Trade of England.* London: James Knowlton.

———. 1701. *An Essay Upon Ways and Means of Supplying the War.* London: Jacob Tonson.

———. 1704. *Essays Upon Peace at Home and War Abroad.* London: James Knapton.

De Grasse, Robert. 1983. *Military Expansion, Economic Decline: The Impact of Military Spending on US Dominic Performance.* Armonk: M.E. Sharpe.

Durbin, Evan Frank Mottram. 1939. *How to Pay for the War: An Essay on the Financing of War.* London: Routledge and Sons.

Edelstein, Michael. 2000. "War and the American Economy in the Twentieth Century". In *The Cambridge Economic History of the United States*, edited by Stanley L. Engerman, and Robert E. Gallman, vol. III. Cambridge: Cambridge University Press.

Einaudi, Luigi. 1927. *La guerra e il sistema tributario italiano.* Bari: Laterza.

———. 1933. *La condotta economica e gli effetti sociali della guerra italiana.* Bari: Laterza.

———. [1943] 1953. "Per una federazione economica europea." In *La guerra e l'unità Europea*, 25–67. Milano: Edizioni di Comunità.
Einzig, Paul. 1934. *The Economy of Re-armament*. London: K. Paul, Trench, Truber & Co.
———. 1939. *Economic Problems of the Next War*. London: Macmillan.
Friedman, Milton, and Jacobson Schwartz Anna. 1963. *A Monetary History of the United States, 1867–1960*. Princeton: Princeton University Press.
Guibert, Jacques-Antoine-Hippolyte Comte De. 1772. *Essai général de tactique*. London: Libraires Associés.
Harvey, David. 2005. *The New Imperialism*. Oxford: University Press.
Higgs, Robert. 1992. "Wartime Prosperity? A Reassessment of the U.S. Economy in the 1940s." *The Journal of Economic History* 52 (1): 41–60.
Hirst, Francis Wrigley. 1911. The Policy and Finance of Modern Armaments. In *Statistical Social Paper*.
———. 1915. *The Political Economy of War*. London: J.M. Dent & Sons.
Jones, John Harry. 1915. *The Economics of War and Conquest: An Examination of Mr. Norman Angell's Economic Doctrines*. London: P. S. King and Son LTD.
Kalecki, Michal. 1947. "Measuring the Cost of the War." In *Studies in War Economics. Prepared at the Oxford University*. Oxford: Basil Blackwell.
Kennedy, Gavin. 1983. *Defence Economics*. London: Duckworth.
Kennedy, Paul. 1993. *Preparing for the Twenty-First Century*. New York: Random House.
Keynes, John Maynard. 1919. *The Economic Consequences of the Peace*. London: Macmillan & Co.
———. 1936. *The General Theory of Employment-Interest and Money*. London: Macmillan & Co.
———. 1940. *How to Pay for the War. A Radical Plan for the Chancellor of the Exchequer*. London: Macmillan & Co.
———. 1978. *Collected Writings*, vol XXII. London: Macmillan.
Kuznets, Simon Smith. 1942. *Uses of National Income in Peace and War*. New York: National Bureau of Economic Research.
———. 1944. *National Product: War and Prewar. Our Economy in War*. New York: National Bureau of Economic Research.
———. 1945. *National Product in Wartime*. New York: National Bureau of Economic Research.
———. 1951. "Government Product and National Income." In *Income and Wealth*, edited by Erik Lundberg, 178–244. Cambridge: Bowes and Bowes.

———. 1961. *Capital in the American Economy, Its Formation and Financing*. Princeton: Princeton University Press.

———. 1964. *Postwar Economic Growth. Four Lectures*. Cambridge, MA: The Belknap Press of Harvard University Press.

Kuznets, Simon Smith, and Raymond Goldsmith. 1952. *Income and Wealth of the United States: Trend and Structure*. Cambridge: Bowes and Bowes.

Lundborg, Louis. 1971. "Statement of the U.S. Senate, Committee on Foreign Relations." In *The War Economy of the United States*, edited by Melman Seymour, 152–160. New York: St. Martin's Press.

Macfarlane, Priscilla. 1937. *The £ s. d. of National Defense*. London and Edinburgh: W. Hodge and Company.

Mann, Michael. 2003. *Incoherent Empire*. London and New York: Verso.

McCarthy, Terence. 1971. "What the Vietnam War Cost." In *The War Economy of the United States*, edited by Melman Seymour, 152–169. New York: St. Martin's Press.

Melman, Seymour (ed.). 1971. *The War Economy of the United States. Readings on Military Industry and Economy*. New York: Saint Martin's Press.

———. 1974. *The Permanent War Economy. American Capitalism in Decline*. New York: Simon & Schuster.

Mises, Ludwig von. [1919] 2006. *Nation, State, and Economy. Contributions to the Politics and History of Our Time*. Indianapolis: Liberty Fund.

Moresco, Emanuel. 1937. *The Colonial Problem: A Report of a Study Group of Members of the Royal Institute of International Affairs*. Oxford: Oxford University Press.

———. 1939. "Colonial Question and Peace." In International Studies Conference on Peaceful Change.

Nordhaus, William, and James Tobin. 1972. *Is Growth Obsolete?* New York: National Bureau of Economic Research.

Pavanelli, Giovanni. 1989. "Finanziamento della guerra e 'circuito dei capitali' in alcune memorie inedite." Parte I. *Storia del Pensiero Economico* (18): 45–70.

———. 1990. "Finanziamento della guerra e 'circuito dei capitali' in alcune memorie inedite." Parte II. *Storia del Pensiero Economico* (19): 61–101.

Pigou, Arthur Cecil. 1916. *The Economy and Finance of the War*. London, Paris, and Toronto: J.M. Dent & Sons.

Robbins, Lionel Charles. 1957. *The Economic Problem in Peace and War*. London: Macmillan & Co.

Rockoff, Hugh, and Geoffrey Mills. 1987. "Compliance with Price Control in the United States and United Kingdom During World War II." *Journal of Economic History* 47: 197–213.

Russett, Bruce Martin. 1971. "The Price of War." In *The War Economy of the United States*, edited by Melman Seymour, 152–160. New York: St. Martin's Press.

Schmidt, Christian. 1991. *Penser la guerre, penser l'économie*. Paris: Édition Odile Jacob.

Shotwell, James Thomson, ed. 1921–1934 vols. 134. *Economic and Social History of the World War*. New Haven: Yale University Press.

———. 1927. *Prefazione del Direttore, a* Luigi Einaudi, *la Guerra e il Sistema tributario Italiano, Storia Economica e Sociale della Guerra Mondiale*. Bari: Laterza.

Skidelsky, Robert. 2005. *John Maynard Keynes 1883–1946, Economist, Philosopher, Statesman*. London: Penguin Books.

Stiglitz, Joseph, and Linda J. Bilmes. 2008. *The Three Trillion Dollar War. The True Cost of the Iraq Conflict*. London: Penguins Books.

# 6

# Economic Rationality or Irrelevance of War?

Classical economists considered war to be an economically mistaken choice, dictated by errors in calculation, or pressed for by particular interests, or even sought for ideal, therefore not rational, motivations. The final decades of the nineteenth century saw the rise of the neoclassical school, which was marginalist and unhistorical, while instead paying attention to calculating utility, equilibria, preferences and optima and being more interested in the economic behaviour of private individuals, families and enterprises than that of the state.

The two World Wars led economists to take up practical problems, such as the organization of war production and finance, price control and inflation, the ways to pay state debt and to pay for reconstruction.

The second post-war period saw a renewed interest in warlike conflict in the theoretical work of some mathematical economists using Game theory to introduce war in their mathematical analysis of the economy, holding that it was a rational activity to the extent that it allowed the minimization of costs in obtaining the objectives it set. As something rational, war could be studied in the realm of conflict through the strategic moves of its actors.



R. Allio, *War in Economic Theories over Time*,
Palgrave Studies in the History of Economic Thought,
https://doi.org/10.1007/978-3-030-39617-6_6

The Public Choice theory was another current that developed in the second post-war period.[1] Its characteristic was to employ economic tools to deal with problems of a political nature. Among its exponents, Gordon Tullock considered more specifically violence, both public and private, starting from the principle that the option of war could be completely rational from an economic point of view.

A still further research path inherent in conflict that followed on directly from the analysis of Smith in the fourth book of *The Wealth of Nations* (2007 [1776]) did not consider war in itself, but rather the economic questions connected with defence, a necessary activity of the state. Christian Schmidt (1991) examined this current and stressed how economists who studied defence differed from marginalists, above all because they believed that economic science should also consider social utility (which was not the same as collective utility intended as the sum of individual utilities), including defence, thinking that social utility did not correspond with the same criteria as individual utility. Social utility, according to what Smith himself had stated, did not accept a convergence between individual pursuit of private interest and the operation of the invisible hand, but saw rather the intervention of the political element "the intelligence of the state" (cfr. Schmidt 1991, p. 75), such that often social utilities were presented as alternatives to individual utilities.

Defence, like justice, was considered a public good, not mainly or exclusively for political reasons, but more correctly for economic motives. Its use was obligatory and indivisible, there was no rivalry in collective consumption, and the benefit for an individual was not a cost for others, or exclusion from access, because prices were zero, because a citizen could be offered defence independently of the fact that he or she had paid taxes or not. The service could be obtained by a further consumer without additional costs, so marginal costs were zero. Defence, the supply of armaments (even in the case of deterrence) and war were entrusted to the state, which in turn decided on the optimization of expenditure on armaments.

The economists who considered "defence economics" in the second post-war period, starting with Charles Hitch and Roland McKean (1960), Edward S. Quade (editor 1964), Alain Enthoven and Wayne

Smith (1971) and Gavin Kennedy (1975), looked into war management as neutral technicians. They did not establish particular economic formalizations, but instead were interested in applied economics, taking into consideration the economic aspects of the problems of defence, with particular attention being paid to the allocation and efficient use of resources for armaments. Their research sought to quantify the weight of the military sector in national economies in order to define the macroeconomic impact of armaments expenditure and to follow its evolution over time. They calculated the cost of defence programmes and evaluated their effectiveness in terms of deterrence, above all when dealing with nuclear armaments. Almost all defence economists also dealt with economic problems that would have been entailed by a political choice of disarmament and, every so often, considered the crux of the relations between military institutions and major armaments manufacturers. Some research was carried out in the direct interest of or commissioned by the armed forces. For example, the book edited by Quade was a collection of the texts of lessons held during courses organized in 1955 and 1959 by the Rand Corporation, a private Californian organization that worked for the United States Army Air Force in producing studies and research on problems of national security. Economic studies on defence today are understandably less numerous than they were in the years of the Cold War, but the research framework remains basically the same.

## War as a Game

Students of military strategy and theoretical economists in the mid-twentieth century began to adopt common tools to analyse the choices made in their respective disciplines and dictated, both in the case of war and in the case of economic competition, by a conflict of interests. The theory used since then for this analysis, Game theory, tried to establish how individuals should behave "under conditions of uncertainty, when such uncertainly derives from decisions unknown to the other individuals interested - generally being opposed to each other - in the outcome of the 'game'". The game could be a war (hot or cold), economic competition or a competition between cunning people (de Finetti 1963, 1, p. 61).

Clearly, this was not the first time that economic choices had been developed through the application of mathematics. It had already been the case with Laplace and his calculation of probability, which could also be used in the study of gambling. Economic theory had thus posed the problem of the rational behaviour of the economic subject and had identified it in the decision that allowed for the maximization of utility/gain with the least effort/investment required. Cournot had applied the principle to the case of monopoly. Later, Walras too had identified a model of general equilibrium, in which, nevertheless, as Morgenstern wrote (1960, p. 125), "every economic subject acts as an individual Robinson Crusoe who is not conscious of his own influence on others and upon whom no influence is brought to bear by others".

At the end of the nineteenth century, Francis Ysidro Edgeworth, an English mathematician and marginalist economist, a contemporary of Jevons and Marshall, proposed the application of mathematical techniques to moral sciences, in particular in order to study individual decisions in conflictual economic situations. In that period, economics tended to analyse the consequences deriving from choices made by agents operating on the market. Edgeworth considered that the rules and calculations that were valid in private exchange could be extended beyond the market sphere to the political struggle for power and trade disputes over wealth too, thereby opening the way for the economic formalization of combat (cfr. Schmidt 1991, p. 69).[2]

The book *Mathematical Psychics*, published in 1881, was not originally intended as an economic tract, but rather as an essay aimed at testing the possibility of applying mathematical methods to sociology through the calculation "of sentiments" in a utilitarian spirit (Edgeworth 1967 [1881], p. IX). In his book, Edgeworth distinguished economic calculation, which analysed equilibrium in a system of hedonistic forces where each person tended to realize the maximum individual utility, from a utilitarian calculation, which studied the equilibrium of a system in which each person tended to realize the maximum general utility and which, despite the apparently unrealistic character, was held to be superior.

Edgeworth inquired into the relationships between wars and economic competition and defined "war" in a general sense of any action undertaken without (or against) the consent of those who suffered the consequences and applied the concept both to economic competition and to war, since, according to him, trade wars between enterprises and political and military wars were only two different forms of the same phenomenon. The analysis of the links between economic calculation and conflict was conducted, for Edgeworth, through the notion of economic settlement in relation to the contract. In the economic field, according to Edgeworth, there had been war before the contract and peace after the contract (Edgeworth 1967 [1881]; cfr. Schmidt 1991, pp. 110). However, seeing that economic calculation itself included war and peace, there was the possibility both to define an economic theory of war and to analyse war in the economic field (cfr. Schmidt 1991, pp. 28, 127).

Two mathematicians took up again the consideration of the hypothesis of mathematical calculation applied to conflict in the interwar period. Emile Borel, in a series of notes published between 1921 and 1927 (cfr. Fréchet 1953; Neumann 1953), and John von Neumann, in 1928, separately conceived Game theory aimed at calculating the best behaviour to adopt in strategy games, that was to say games of intellectual ability in which, nevertheless, there was always a component of chance. Borel, starting in 1921, held that the mathematical analysis of the behaviour of players in strategy games could have applications in the field of economic, financial and military problems, but failed to establish a basic mathematical theorem. Neumann instead was able to do so with the Minimax theorem, a method for minimizing the maximum possible loss or maximizing the minimum gain in zero-sum games, in which one player's gain was another player's loss.

All the same, Borel and Neumann's writings did not have a wide distribution outside the community of students of mathematical science. The theory emerged from this circle and was popularized towards the end of the Second World War thanks to the book *Theory of Games and Economic Behavior* that Neumann wrote in 1944 with Oskar Morgenstern, an exponent of the Austrian School of economics and author later

(1959) of the hard-hitting and sometimes apocalyptic *The Question of National Defense.*[3]

Game theory intended to deal with economic problems in the presence of competitive situations that could not be considered by traditional mathematical formulae, even if the book by Neumann and Morgenstern was short of true economic examples. The first applications of the theory to the science of warfare were during the Second World War and then during the Cold War.

John Nash, a mathematician and economist, Nobel Prize winner, contributed to opening the way for the application of Game theory in economics by defending the situation of equilibrium that occurred in non-cooperative games in the presence of interdependent strategies, that was when the behaviour of a player depended on the choices made by the other. Nash's equilibrium was reached when the behaviour adopted by each player was the best possible response (dominant strategy) to the choices made by the adversary. However, Nash also demonstrated that this rational behaviour, that led to realizing the best individual result possible (individual optimum), determined instead a non-optimal situation at a social level. Therefore, it demonstrated that the social optimum was not compatible with rationally selfish individual behaviour. This meant that the existence of Smith's providential invisible hand was therefore denied.

Starting in the 1960s, Game theory began to be applied to various disciplines outside economics and war strategy, including biology, psychology and psychiatry.

The characteristics of strategy games were summarized by Morgenstern in a book published in 1960: every player wants to win the most; the quality of information is often limited; the cards are distributed randomly; he has to realize that the other players will respond to his moves and will try to discover his intentions just as he would like to discover what the others intend to do. The final outcome does not depend on any player taken alone, but on all taken together because each player controlled only a part of the variables that together determined the result (Morgenstern 1960, p. 85). These conditions were well adapted to the study of strategy, both military and economic, leaving aside the fact that economic processes, for example trade, were not zero sum because in trade all gain, or at least hope to do so.

Further developments included those of going on from zero-sum games with two players and without cooperation to "n" players in non-zero-sum games with cooperation.

Nevertheless, the validity of the concept of rationality, which should determine choice, attracted many well-grounded doubts, as regards both the economic applications and warlike ones. In particular, in the collection of Papers in Honour of Hirschman, edited by Foxley, McPherson and O'Donnel, Amartya Sen and Alessandro Pizzorno (1986) dedicated their papers just to this problem. Sen concluded his contribution by stressing "the opportunity to resist what Hirschman has called 'the progressive impoverishment of the prevailing concept of human nature'", adding that rationality must offer this concept greater space (Sen 1986, p. 351).

Lewis Fry Richardson, a physicist by training, mathematician, meteorologist, student of psychology and English pacifist, analysed the causes of international conflict and the ways to avoid them, writing at the end of the Second World War in the same period in which Game theory was being developed. He did not use Game theory, but instead a system of differential equations with the calculation of probability. He widely used historical series of economic data, such as the amount of resources a country dedicated to armaments (an index of likeliness of war and of the current level of tension) and the trend of world trade, used as a measure of cooperation between nations, therefore implicitly attributing a pacifying function to trade.

Richardson took into consideration the period 1815–1945 and analysed the distribution of wars and their intensity over time, measuring intensity on the basis of the number of wars taking place with the number of dead, concluding that there was no proof that during the period there had been a net tendency towards an increase in the number or frequency of wars, while the dimension did increase (1950 manuscript).

Quincy Wright reached the same conclusion in his 1942 book *Study of War*. Years later (1980), Michael Mann used the historical series complied by Sorokin (1937) and by Singer and Small (1972) and confirmed the tendency, as far as Great Britain was concerned, for the very long period 1130–1815.

Richardson drew on the analyses of various historians to identify the economic causes that had unleashed wars, or which could have been considered responsible or co-responsible for domestic and foreign conflict, not failing to note the difficulty in making a rigorous definition of the field of clearly economic causes. These causes often were seen to be closely correlated with a cause of a different nature, such as national pride, which tended to increase the perception of economic damage deriving from the trial of strength with hostile nations. Richardson further added that one could always say that every human question was economic, as humanity depended everywhere on material goods, such as food, land and equipment; but then we could by analogy state that every human question is also sexual for the simple fact that we have all been born (Richardson [1950] manuscript 1960a, p. 210).

Stating this, in his major works published after his death: *Arms and Insecurity* (1947–1949 manuscript) and *Statistics of Deadly Quarrels* (1950 manuscript),[4] Richardson tried to analyse statistically the causes of war and judged that 29% of conflicts that broke out after 1920 seemed to have a direct economic cause. Economic causes were more largely present in determining "small" wars than greater ones. Disputes and trade restrictions had caused wars just like the wish to acquire new territory to strengthen trade, invest capital or settle colonists. The excessive taxation of colonials or domestic minorities and economic assistance to the enemy were further causes, while major income disparities, contrary to what Marxists presumed, seemed to have had little importance in unleashing war, at least for the period under consideration.

Richardson also elaborated a theory based on the idea that the willingness of neighbouring nations to go to war was a function of the length of their common border and constructed a model of the international dynamic of military and armaments expenditure. Years later, some of his considerations were adopted by Anatol Rapoport.

Richardson, along with the politologist and student of international relations, Quincy Wright, was considered to be among the principal pioneers in the vast field of conflict analysis. In 1960, Richardson described the economic instability of states that based their security on deterrence, investing huge sums every year in unproductive goods such as armaments, thereby limiting the possibility for investments that could produce new wealth.

Wright, in his monumental and passionate work *A Study of War* (1942), begun in 1926 and completed fifteen years later, studied war in order to discover the conditions under which peace could be realized as an equilibrium between various different forces, but making no use of mathematics. He collected facts and judgements on the war, concluding that war could draw its origins from technology, particularly from its military applications; from law, especially when it regarded the norms of conflict; from social organization; and from common opinion on basic values. He also dealt with war in subsequent publications: *The World Community*, published in 1948, and *The Study of International Relations* (1955).

Anatol Rapoport, a Russian mathematician and psychologist, a naturalized American citizen, posed the problem in the introduction to his book *Fights, Games and Debates* (1961, pp. xii–xiii) of the validity of the use of Game theory as an analytical instrument in various disciplines and stressed how "humanists" could quite rightly accuse the theory of overlooking ethical problems; sociologists could criticize it for not having taken the social structure into consideration in dealing with the formation of coalitions, while for psychologists the assumption of "complete rationality", typical of the theory, was simply unrealistic. Rapoport did not mention criticisms from the point of view of economists, but stressed that nevertheless there was a considerable difference between establishing a priori the existence of a shortcoming within a theory and coming to identify this shortcoming through reasoning. In the former case, the conclusion could be reached that the theory was not correlated with the argument dealt with, in the latter case that a way could be found to make the theory suit the subject to be analysed. This reflection took place because Game theory had had a very strong impact on his conscience and his way of thinking about conflict. In 1960, Rapoport recalled that about ten years beforehand he thought that conflicts derived principally from the heightening of the clash between different points of view, and therefore, it was always possible to resolve them. At that time, the idea that war could be seen and analysed as if it were a game horrified him.

According to him, it could not be just a game when it was a matter of the well-being or the very lives of people at stake. He further held that the principal result of the war "game", power, was a perversion of the objectives of human life.

He later concluded with the analysis of Game theory that if ethics were set aside for the moment, looking at the theory for what it was, and if, at the same time, the hypothesis of the existence of irreconcilable conflicts could be accepted, then the "logic" of situations under examination could be followed and a different insight cast on conflict and the opportunity for the development of each discipline that was proposed to deal with the problem scientifically. Any type of scientific approach could then be used to clarify the reasons for the conflicts and how they were conducted or came about. Two authors introduced it to the problem of the ethical implications of Game theory: Braithwaite, who stressed the positive contribution of Game theory in the examination of ethical problems in his essay *Theory of Games as a Tool for the Moral Philosopher* (1955), and Schelling, who "launches a spirited attack" on this idea (Rapoport 1961, pp. xii–xiii).

Nevertheless, in the end, Rapoport maintained his opinion that if Game theory could be used to understand the complexity of situations which strategies implied, it was not able to indicate the best way to behave on different occasions. On the contrary, it made clear that a single "solution" often did not exist, that sometimes the games turned out to be "badly designed" and it was not even possible to find a reasonable solution in the sense that there was not just any way to participate "*reasonably* in the game, a way when whoever acted according to criteria gained an advantage in reaching his or her objectives" (de Finetti 1963, p. 69). He therefore advised against using games designed as "suicide traps" and instead proposed opting, in determinate circumstances, for trust and reciprocal comprehension.

Therefore, "the value of Game theory is not in the specific solutions it offers in highly simplified and idealized situations, which may occur in formalized games but hardly ever do in real life. Rather, the prime value of the theory is that it lays bare the different kinds of reasoning that apply in different kinds of conflict" (Rapoport 1962, p. 114).

Rapoport showed above all how the game of poker in the case of the application of war decisions was neither the more general nor the most sophisticated model of conflict and not even the most meaningful for applications and stressed how it would be difficult for a system based on selfishness (i.e. the "game" of weighing up hopes of gains or prizes against fears of losses and punishments) to lead to a common good, and even if the system functioned, it would be morally repulsive because authority would be based solely or mainly on threat or fear of punishment, or alternatively on the attraction of gains, not effectively drawing human beings towards the common good, and even if we supposed it did draw them, it would not be in line with their dignity as people, that is, as rational and free beings. These arguments were also accepted by de Finetti (cfr. de Finetti 1963, 5, pp. 70–71).

Thanks to his studies in psychology, Rapoport shows a particular sensibility to the criticisms of the supposed scientific rationality which drives "the mathematics of strategy". During the Cold War, when he was revolted by the proposal of the strategists financed by the Pentagon to "think the unthinkable" and to calculate "the acceptable cost" of a nuclear war in terms of millions of human lives, Rapoport proposed the confutation of the scientific validity of the mathematical calculations forming the basis of such hypotheses. His book *Strategy and conscience*, published in 1964, sustained the falsity of the principle of absolute objectivity in scientific thought because the points of view and the subjective opinions of researchers are an essential part of scientific analysis.

> The logician, the mathematician, the statistician, and the strategist all derive their competence (and so their social status) from an ability to handle abstract chains of reasoning detached from content. This, rather than freedom from preconceived notion and from the bias of vested interest, is the true meaning of their detachment. (Rapoport 1964, p. 192)

For example, in the case of Game theory applied to military strategy, the estimates of probability that the strategists worked with had a largely subjective character. Furthermore, the simplification of the data for the situations under examination that required a mathematical command of the problems led to formulae that were inadequate in representing the complexity of the real situation. Lastly, Rapoport revealed the ignorance of these mathematical analysts of the real reasons for the behaviour of people, which were dictated much more by their psychological make-up than by the number of prizes or successes obtained.

To the mathematician among them, equations on the blackboard are just equations. Mathematics is a great leveller. When a problem is mathematically formulated, its *content* had disappeared and only the form has remained. To the strategist "targets" are indeed only circles on maps; overkill is a coefficient; nuclear capacity is a concept akin to heat capacity or electric potential or the credit standing of a concern. The logic or abstract reasoning applies in the same way to all problems which are logically isomorphic" (Rapoport 1964, p. 192).

Rapoport was not the only one who posed this type of problem, as other mathematicians too manifested their unease in applying them to the case of war after having dealt with the theme of decisions in a conflictual situation from a logical-theoretical point of view. This was not the case for Neumann, the founding father of that theory, who, even if the title of his first widely read book, *Theory of Games and Economic Behavior,* referred to economic applications, linked it equally rapidly to war.

Von Neumann, Hungarian born, but naturalized an American, was in fact a member of the United States Atomic Energy Commission and had been involved in the Manhattan Project for the construction of the atom bomb. He had been very active in calculating and suggesting the best way to launch bombs in order to obtain the highest level of deaths and damage. In terms of conflict, Neumann supported the opportunity for the United States to equip itself with a strong nuclear arms contingent and was even more eager and incisive than McCarthy in suggesting the pre-emptive bombardment of the Soviet Union to the American government in order to avoid the spread of communism.

More generally speaking, when analysing conflicts, Game theory was used right from the start to identify and evaluate the war situations that could occur following certain decisions. The sociologist Raymond Aron appreciated the usefulness of this application, because it allowed the dialectic of hostility to be posed in an abstract form, so avoiding taking decisions based on of forecasts for an unknown future instead of based on the approximate frequency known to us of various types of event (Aron 1970, p. 879).

Morgenstern (1959, pp. 263–265) had cause to complain, at the end of the 1950s, about the lack of attention paid, above all by politologists, to the use of Game theory, while in 1963 de Finetti noted that only Martin Shubik (1959), an economist and Professor of Mathematical Institutional Economics at the University of Yale, had written a book on the applications of Game theory to economics, above all to the problems of imperfect competition.

The situation changed rapidly in the following decades, and a collection of essays edited in 1987 by Gerard Radnitzky and Peter Bernholz stressed the intrusiveness in that period of quantitative economic methods in the analysis of problems inherent in various disciplines. Here, Jack Hirshleifer, an Economics Professor at the University of California, dealt with the economic-mathematical approach to conflicts, in particular those leading to the use of violence, and held that rational choice models at the level of decision-making and equilibrium models at the level of social interaction could be most useful in studying them. However, Hirshleifer also held that conflict analysis could, in turn, offer much for the study of the economy "Attending to the darker aspects of how humans might and do compete is absolutely essential even for a proper understanding of the relatively benign nature of maker competition" (Hirshleifer 1987, 1, p. 335). For him, conflicts were a kind of "industry", a way by which economic agents weighed each other up in terms of resources and asked what circumstances pushed the parties into conflict if conflict was not always, or even often, an error on the part of one of the contenders, if not all of them, and if better information could not promote pacific agreements.

The economist, without being a manager or an engineer, could apply certain principles to an industrial production process and offer solutions. So too, without wanting to substitute military command, the economist could be able to say something useful on the possibility of obtaining the desired results through violent conflict.

Years later, Hirshleifer returned to the dark side of human behaviour. In one of the numerous essays edited in the collection *The Dark Side of the Force* (2001), Hirshleifer defined himself as a student of Schumpeter and considered predatory modes and conflicts rather than like the majority of economists, who investigated the production and exchange of wealth. He did so starting with a quotation from Pareto recalling how the efforts of man to obtain wealth can be used in two different ways: to produce or process goods, or to appropriate the goods produced by others. Hirshleifer held that wealth was often acquired individually or socially in the second way, also because in a world that nevertheless required the organization of defence against the aggressor, even politicians who pended to pacifism had to balance their decisions between the two strategies and, in reality, all choices were made "under the shadow of conflict".

Hirshleifer did not mention the obvious fact that to be able to plunder, someone else had to continue to produce, but instead showed that predatory methods were "economic" too. In fact, when responding to the ever present factor of scarcity, these methods for acquiring wealth left the door open for rational choices and in these cases "decentralized" decisions interacted to achieve a social equilibrium (Hirshleifer 2001, p. 2).

Production and exchange tend to increase social wealth. Conflicts and predatory methods, while appropriation and sequestration were limited to its redistribution, obviously excluded the part that was lost in the fighting. Hirshleifer included in the dark side of force crime, war and politics, that is to say everything the tradition of Marshall did not take into consideration and economists in general overlook, even though more recently Public Choice economists have begun to examine the struggle for the control of resources, under the name of "rent-seeking".

If he could admit that biological motives of food and sex that caused conflict to break out in ancient society had been superseded in contemporary society, Hirshleifer held that economic motives were not alone in arousing contemporary conflicts among which intangible aspects stood out too, such as prestige, domination and respect. These latter motives were originally individual, but later transferred to the social level of groups with a common identity in religion, ideology, culture and nationalism. Here, in fact, we can see the reference to Schumpeter.

The causes of war and peace were, in the end, the same for Hirshleifer. War and peace were alternative strategic choices aimed at reaching the same goals. The choice fell on one option or the other following contingent factors, dictated mainly by opportunity and different perceptions (Hirshleifer 2001, p. 39).

Hirshleifer listed the contributions in his introduction to the *Dark Side of the Force* of those, starting with Schelling (*Strategy of Conflict*, 1960), who had in their different ways dealt with the theme of economic rationality in the use of violence. Apart from the essays by Boulding (1962); Tulloch (1974); Bernholz (1985); Usher (1992), Hirshleifer also recalled many contributions to scientific journals.

An interesting critical summary of mathematical analyses of a neoclassical nature above all from the 1990s on the theme of war was published in 2002 by Christopher Cramer in "World development". Cramer recognized the merit of these analyses, in particular those conducted using Game theory, in providing a material explanation of conflicts and of having implicitly introduced the economic element to the motivations for war, something not even taken into consideration in other kinds of analysis. His entire criticism from the point of view of the economic policy of war nevertheless comes down hard. Cramer in fact holds that the theory of rational choice, based on the precepts of neoclassical economics, when applied to conflict, ceased to be convincing from a theoretical point of view and when an empirical content was presented, it was often selected in a highly arbitrary manner. The economic explanations of the violent conflicts accepted by this theory appear for him to be "extremely reductionist, highly speculative, and profoundly misleading" (Cramer 2002, p. 1849). Alternative approaches in the field of economic policy could, he considered, provide a more realistic economic content. Cramer stressed how the choice between conflict and cooperation was not taken in reality solely on the basis of economic utility, as conflicts could be institutionalized and peoples could be mobilized by ideological factors or by promises of change. Another incongruence derived from the fact that a social phenomenon like war was investigated through the use of models based on rational choices of an individual kind, "and on a very narrow notion of rationality" (Cramer 2002, p. 1850). Lastly, even if orthodox economic theories and econometric applications could succeed in grasping a part of the reality of conflict, they failed totally in capturing one of the most typical characteristics of all conflicts, the structural and relational changes that derive from them.

Cramer recalled that, except for Keynes, economics arrived rather late in examining conflicts and that economic theory until the 1990s had not paid much attention to the problem of war. Since then, the studies cited on models of conflict, based on methodological individualism and rational choice, have begun to proliferate.

The applications of Game theory to economic choice and decisions on war have markedly interwoven over the years to the point that in 2005 the Nobel Prize for economics went to two mathematicians who were interested mostly in the application of Game theory to war strategy and in their acceptance speech spoke of war, not economics.

## Robert Aumann, Gordon Tullock and Rational War

Robert Aumann won the 2005 Nobel Prize for economics, along with Thomas Schelling for "having increased our understanding of conflict and cooperation through the analysis of game theory."[5]

Aumann, and Schelling too, is not an economist, but a mathematician and analysed warlike conflicts much more than economic ones using Game theory. During an interview with Sergiu Hart (2005), Aumann recalled that he started to be interested in Game theory during his period of post-doctorate studies in operations research at Princeton. The project he was working on at the time was sponsored by Bell Laboratories, which was developing a missile defence system.

The speech he made on his acceptance of the Nobel Prize bore the title *War and Peace.* In it, Aumann stressed that in history nothing has been more constant than war which, as a *phenomenon,* was not at all a series of isolated events and could be studied in its general characteristics. Research could be conducted from an historical, psychological or rational point of view. The economic point of view was not mentioned.

In his dissertation, Aumann stated he wanted to answer the question, "why does *Homo oeconomicus* –rational man – go to war?" and even though on other occasions he had sustained that rational man belonged to a mythological species and that *Homo sapiens* acted most of the time for unconscious motives, or conscious but irrational and emotive ones too, in this case specified that rationality was for him as defined by Walras according to whom "a person's behavior is **rational** if it is in **his** best interest, given **his** information". Starting from this definition, Aumann therefore held that war could be rational, and as such could be understood and analysed in its general characteristics, in its common aspects and in the differences it presented on different occasions. If instead it was considered as irrational, the problem could not be dealt with by his analysis.

In fact, war appears to be the most irrational of human activities when seen from the position of those engaged in its combat, or suffering its effects, or humanity in general. This is not the case if it is viewed from the position of the politicians who declared it and the economic interests that required it. War, for them, can be rational in the economic sense expounded by Walras. Even if since the last century, almost always those who unleashed war turned out as the losers. But even in these cases, the consequences for politicians were not serious. An empire was not lost due to Waterloo. The terrible disaster in Iraq did not stop Bush Jr. from completing his term in office, and Tony Blair got by when he admitted he had made a mistake.

Aumann continued his reasoning by offering up two other quotations, one by Abraham Lincoln: "both parties deprecated war; but one would make war rather than let the nation survive; and the other would accept war rather than let it perish. And the war came" (Aumann 2008a, p. 351). The second quotation was from Jim Tobin, another Nobel Prize winner for economics, according to whom economics was summed up in just one word: "incentives".

After declaring that he did not want to talk about how war was financed or about how the subsequent reconstruction was carried out, Aumann focused on the incentives that led to the outbreak of war and on ways for creating incentives that could prevent it instead.

The economy too could be considered as a game in which the incentives to the players interacted in a complex and surprising way. Aumann used an example from another field, taxation, to illustrate the concept. He thought if someone wanted to increase income from taxation, it would be opportune to lower, not to raise, the percentage of taxation and at the same time provide incentives to work, reduce tax evasion and stimulate the economy. In the same way, if someone wanted to prevent war, armaments should be increased. Clausewitz had said the same thing, but Aumann did not quote him. Instead, he used a more recent example, the Cold War, when nuclear bombardment could be carried out at any moment by either of the two leading powers, who therefore found themselves to be in the position of dissuaders and dissuaded at the same time. It was just this fact that had made the maintenance of peace possible. It was MAD, Mutually Assured Destruction, that guaranteed the maintenance of the strategic equilibrium in the repeated games of the dilemma of the prisoner in a warlike version.

"If you want peace now, you may well never get peace. But if you have time – if you can wait – that changes the whole picture; *then* you may get peace now". It's one of those paradoxical, upside-down insights of Game theory, and indeed of much of science (Aumann 2008a, p. 355).

If Rational Choice Theory, largely accepted by contemporary economists,[6] was valid also for decisions to declare war, it would be logical to think that the states would carefully evaluate costs and benefits, the number of victims, military expenditure, possibility of victory and length of engagement before deciding on the basis of the maximization of advantages. An authority on the matter like von Clausewitz warned that war was in the hands of fortune and in war the unexpected was always met. The conflicts over the past decades have proved him right. Those who have unleashed war have almost always lost or, in any case, achieved a result different from the one predicted. Leaving aside the two World Wars, this holds true for the Korean War, the Suez Crisis, for the attacks on Afghanistan, firstly Soviet and then American, and for the Gulf Wars.

Public Choice, a theory we shall come back to later, agreed to consider violence and war as possible areas to study with economics. Gordon Tullock, the founder of this current of thought with Buchanan, was interested above all in the theme of war and judged war to be rational, as any action is rational if it permits reaching planned objectives while minimizing the costs meeting them, independently of any moral judgement on means and ends. "Unfortunate practices" under certain circumstances can in fact offer advantages to some, and Tullock clearly said that he did not deplore them at all. Stating that violence could be rational may seem to many people a contradiction in terms, but it is not so. Violence, coercion and war are neutral words in themselves and do not have any particular connotations of good or bad, and if they are desirable or undesirable can be decided case by case.

Tullock used a reasoning similar to the one the mercantilists had used before when he sustained that violence can be highly desirable for a subject (public or private) not because the subject liked violence in itself, but because it hoped to gain an advantage from it. Violence and war constituted a consistent part of human history and war was conducted either to make a gain or to avoid a loss (Tullock 1974, pp. 1–6). Even those who declared they did not want to use violence, if they fell victim to plunder, would normally have used it to react.

The conception of "modern" economics was that profit was reciprocal in transactions and that economics traditionally studied the benefits of cooperation. Nevertheless, the distribution of profits among the parties could be the object of conflicts, and even if conflicts consumed resources, and consequently were not socially efficient, entering into conflict could be individually useful and rational for one or both parties engaged in it.

Tullock also paid interest to the problem of minimizing costs of conflicts and held that war was the best example of economies of scale, because in general the increase of resources destined for combat brought a more than proportional increase in the probability of success. All this was represented on Cartesian axes where, however, as explicitly stated, the possibility of victory was calculated, not the fact that thanks to victory one could be better off than before (Tullock 1974, pp. 89–90).

As regards nuclear war, Tullock recalled that some people considered it as an impossible alternative. Whoever supported it implicitly assumed that this type of conflict would have meant a level of damage too high even for the victors. Tullock admitted that this could have been true, in particular in the 1970s in the case of a war between the United States and the Soviet Union, but this argument did not appear generalizable to him. Nuclear war could have constituted an alternative (Tullock 1974,

p. 88), and atomic arms of a great power could have functioned as a deterrent and guaranteed unarmed allied countries too against attacks from a hostile nuclear power.[7]

As regards the controversies that took place in domestic transfers of wealth, the investment of resources in war was often rational for the individual who intended to gain wealth at the cost to others, or who saw the necessity to defend their own wealth, but this type of behaviour constituted a net waste for society because it implied the need for the state to set up a public force able to control and reduce the costs of conflicts. The organization of the public force would clearly have a cost.

There was no way to effectively control international conflicts. Tullock did not accept enlightenment thought on the perfectibility of human institutions; in particular, he did not believe in the perfectibility of public institutions, which nevertheless could be improved. If the dreams and utopias of a world freed from violence were abandoned, it would have been possible to achieve the realization of a world which certainly would not be perfect, but which would be better than the present one (Tullock 1974, pp. 139–140). This conclusion is substantially in common with the thought of Aumann and Schelling.

## Thomas Schelling and the Power of Deterrence

Schelling, who shared the Nobel Prize for economics with Aumann, awarded to him for the same reasons, read a speech at the prize-giving ceremony dedicated to the power of deterrence against the outbreak of war exercised by huge armaments, above all nuclear, whose use had become a taboo after Hiroshima.[8]

Schelling followed Aumann in carrying out his analysis of conflicts through variable sum Game theory in which "the sum of the gains of the participants involved is not fixed so that more for one inexorably means less for the other. There is a common interest in reaching outcomes that are mutually advantageous".

Schelling thought that Game theory found a privileged application in the doctrine of dissuasion resting on a system in which the decisions of each actor depended on the forecasts of the intentions of the others and on the anticipation that the others could make of these decisions. This could give rise to places and forms of interaction/cooperation between players.

Schelling was interested in deterrence starting from the basic principle in which "to study the strategy of conflict is to take the view that most conflict situations are essentially *bargaining* situations", which could therefore be resolved by a process of negotiation. "The bargaining may be explicit, as when one offers a concession; or it may be tacit maneuver, as when one occupies or evacuates strategic territory". In turn "viewing conflict behavior as a bargain process is useful in keeping us from becoming exclusively preoccupied either with the conflict or with the common interest" (Schelling [1960] 1980, pp. 5–6). Schelling adds that "Though 'strategy of conflict' sounds cold-blooded, the theory is not concerned with the efficient *application* of violence, or anything of the sort; it is not essentially a theory of aggression or of resistance or of war. *Threats* of war, yes, or threats of anything else; but it is the employment of threats, or the threats and promises, or more generally of the conditioning of one's own behavior in the behavior of others, that the theory is about" (Schelling [1960] 1980, p. 15).

Reflections on the dissuasive capacity of nuclear armaments began in the United States after it lost its atomic monopoly in 1949. In 1960, on the publication of the book on conflict strategy, Schelling declared that, as regards war, studies on deterrence that had been published up till then, including academic ones, had been essentially concerned with resolving contingent problems rather than seeking to accumulate information and generate discussion aimed at being able to develop a theoretical structure. Consequently, international strategy appeared to be "a science running late". When the volume was republished in 1980, he recognized that the gaps had been filled.

A presupposition of Schelling's studies was that there are similarities between economic strategies and military ones, if only because in both cases there is a situation of a conflict of interest. This allowed for the use of common tools for theoretical analysis, naturally Game theory above all. In the beginning in this field, the economy was generally understood not as a zero-sum game, because all the players taking part thought they would win, while war was presented as a zero-sum game, where the winner took all and the loser lost, following the beliefs once enunciated by the mercantilists. Schelling, however, showed that war should be considered as a non-zero-sum game, seeing it was negative for all the participants. That meant that all participants lost,[9] which was an additional motive for following a policy of deterrence.

The economy seemed to be similar to a cooperative game and war to a non-cooperative one, but it was not so in reality: trade competition was not at all cooperative because each agent acted alone, as Cournot had shown in 1838 in his book *Recherches sur les principes mathématiques de la théorie des richesses.* When dealing with the competition between two producers, Cournot had in fact written that "*chacun de son coté*" (each for his own) sought to maximize profit, adding that this specification was essential, because if the producers agreed to both obtaining the greatest possible profit, the results would have been different and prices would have been those of a monopoly (Cournot [1838] 1938, pp. 88–89, cfr. Le Bras-Chopard 1994, pp. 111–112).

As regards war, on the other hand, it did not always have to be the case that the interests of the adversaries were always conflictual. In certain circumstances, they could have had common interests, which during a conflict could have led to forms of cooperation, for example, to reduce the intensity of the fighting or at least to find solutions that were less devastating for all concerned. Schelling recalled that tacit negotiations were not rare between belligerents, as was the case for the Korean War, which remained a limited conflict, thanks to cooperative-type agreements between the Soviet Union and the United States aimed at preventing the conflict spreading further and causing more damage. The Cuban Missile Crisis, which took place the year after the publication of the book, would have offered another important example of cooperative agreements.

War seemed to regard qualitative rather than quantitative games. Then, if economic activity could be seen as similar to repeated games, while in war situations, when thinking in terms of battles, it would be difficult to see them as the repeat of a game, given the extreme variety of cases that can occur during combat.

Many criticisms have been made of these theories, some coming from mathematical economists that demonstrated the irrationality of the equilibrium obtained through dissuasion (Selten 1978), and others, which are widely known, consider the fact that the logic which rules economic behaviour is not at all the same as the one that rules military strategy. In the field of the economy, the aim is to obtain the best result at the lowest cost, and in war, the aim is to submit the enemy. Military strategy has more differentiated and complex objectives than the objective of the production of goods and services.

As for the rationality of the behaviour of the actors in the field, Schelling himself posed the problem:

> Furthermore, theory that is based on the assumption that the participants coolly and 'rationally' calculate their advantages according to a consistent value system forces us to think more thoroughly about the meaning of 'irrationality' [...]. Rationality is a collection of attributes, and departures from complete rationality may be in many different direction. (Schelling [1960] 1980, p. 16)

Luttwak, the author of several books on military strategy, took up von Clausewitz's considerations in a book published in 1987 to stress how economic logic diverges from the logic which rules choices made in war, while marginal calculation of costs and the goal of economies of scale were important in the definition of economic effectiveness and did not appear to be applicable to the field of war strategy, where surprise attacks and "paradoxical" choices had a completely different importance. Nevertheless, Luttwak was convinced, like Schelling, of the power of deterrence in the arms race. His *On the Meaning of Victory*, published in 1986, maintains that when war is to be considered as totally improbable, the conditions to unleash it are already in place for two reasons: firstly, since war is considered as an absurd eventuality, we refuse to pay the price of deterrence by maintaining large-scale armed forces. Secondly, for the same reason, we refuse to make the diplomatic concessions and political compromises required to eliminate the incentives of going to war (Luttwak 1986, p. 172).

The technique of dissuasion through the build-up of armaments is not, however, an exclusively strategic problem, as it also involves the economic dimension because it presupposes the ability to deploy modern and effective armaments that are necessarily expensive. Luttwak admits that deterrence is the strategy adopted by strong countries—countries that are so strong that they can show a potential enemy that their defence capability is such that any armed attack would only be successful at the cost of enormous losses, or even the total destruction of both parties. Economically weaker countries, if they did not form an alliance with stronger ones, would have no choice but to take recourse to attempting offensives, stratagems, improvisation and simulation.

If we return to Game theory, following the publication of Schelling's books, economists began to pay greater attention to the theory, including in their university teaching.

Twenty years ago, Armelle Le Bras-Chopard made a short comment on the studies of Schelling and the application of Game theory to economic and warlike conflict, reaching the conclusion that I feel can still be shared. "Useful as they are in demonstrating certain correlations between the orders of economic and strategic facts, these instruments, which represent economic models, offer limited results and do not allow us to reach a general explanation of war" (Le Bras-Chopard 1994, p. 113).

In 1962, Kenneth E. Boulding, Professor of Economics at the University of Michigan, analysed conflicts from another point of view. He was convinced that war was the main moral and intellectual problem of the time and was equally convinced of the inadequacy of the pacifist movement in confronting it. He published a book on conflicts and defence, which, as the author immediately made clear, was a completely theoretical work that used mathematical instruments, including Game theory, aiming to go deeper into the study of conflicts understood as social processes, war just being a special case of them. Boulding coined the term *irénologie* to designate the science of peace which was concerned with negotiated solutions to conflicts as opposed to *polémologie* proposed by Gaston Bouthoul to mean the science of war. *Irénologie* did not meet with a great success as a term, but Boulding's studies under this name were the same as what is now called "peace research", or "conflict resolution". Boulding and a small group of students who agreed with his ideas founded a periodical called "Journal of Conflict resolution" in 1957 and two years later set up the Centre for Research on Conflict Resolution at the University of Michigan.

Boulding, in his book *Conflict and Defense*, intended to make his own personal contribution to the understanding of problems connected with peace and war by applying his own theoretical construction, based largely on the economic theory of oligopoly, albeit with contributions from various sciences, such as sociology, psychology and ecology, to various fields, including international relations and conflicts. In his analysis, Boulding extended the examination of changes that war caused on the economic structure (price rises, movement of wealth and investment from one sector to another) in the period of rapid rearmament preceding the outbreak of hostilities and then again in the subsequent phase of reconstruction during which the economic consequences of the conflict were measured and the relations of force were reorganized.

On concluding the theoretical part of the analysis, Boulding nevertheless ended the volume by recognizing that conflicts had practical implications, including survival, which he wished to examine, even though this meant resorting to ethical judgements. He therefore dedicated the last two chapters to research into the ways of controlling and resolving conflicts, paying special attention to warlike conflicts, stressing that negotiation always remained one of the ways to find a solution to conflicts. He made an effort to distinguish between the dynamics typical of the war system and those of the diplomatic system.

## The Disinterest of the Neoclassical Economists

Theoretical economists of the twentieth century, except for students of Game theory, concentrated and still concentrate on situations of peace and the behaviour of private citizens. The difficulty met in applying neoclassical theory to war derived from the imperfection of wartime markets, from the difficulty in accessing information and the dichotomies of the positive externality of the defence of a country and the negative effects on its enemies. Having stated this, the doctrine has also often ignored the study of the incidence and the effects of military expenditure on the civil economy, which are quite great. These researchers propose the identification of the optimal solutions for determinate economic problems starting from given conditions which exclude disturbing elements.

In 1983, Christian Schmidt, Professor at the University of Paris, presented a study carried out for the United Nations on the consequences of the arms race, noting that in consideration of a rapid increase in the production and world commerce in armaments, some associations of economists[10] had requested their members to look into the problem of economic fallout. This request did not receive any replies, seeing that the amount of research conducted on the subject was slight, marginal and fragmentary (Schmidt 1983, p. 3). Schmidt put the causes down to the difficulty of getting on top of the argument, apparently still so vaguely outlined, and of gaining access to the specialization in studies which led to considering questions relative to armaments and war as belonging to the field of politics.

That state investment in armaments and the proliferation of wars had not fallen in worldwide importance over the years was forcefully stressed by Alan Milward, a student of the European war economy, when he noted that nevertheless, and slightly incongruently, peace was generally considered by economists "the state of affairs most conducive to the achievement of economic aims and the one which economics theory seeks to analyse and illuminate". Not only because at the beginning of the twentieth century peace had in fact been assumed to be the goal economic theory should move towards (Milward 1977, p. 2).[11]

According to Milward, the frequency of wars in contemporary society should have instead led students to consider war as a state of normality, thus making its economic manifestations an object of analysis. However, the majority of economists continued to fail to face reality and proceeded to construct theoretical models tending to equilibrium, a situation which was obviously incompatible with the reality of war. Jack Hirshleifer (2001, p. 3), for his part, noted that the term *war* did not appear in the index of Marshall's *Principles of Economics*

Back in 1942, Quincy Wright ([1942] 1965, pp. 708–719) had stressed that the majority of economists, excluding the mercantilists and the Marxists, had considered war as an element extraneous to their own research interests. Nevertheless, he drew attention to the fact that some economists of a liberal background had been forced to ask themselves at least why war existed, following the theses put forward by Marxist analysts and by historians and students of international relations, who had bit by bit identified the reasons for warlike conflict in capitalism, imperialism, the international arms trade and big international finance. Neoclassical economists who had asked themselves the question had generally concluded that both economic theory and historical evidence had demonstrated the marginal importance of those reasons. For example, Lionel Robbins (1939) and Eugene Staley (1935) had expressed that kind of opinion.

Thorstein Veblen (1917) instead linked capitalism with war. In general, it was not economics students who insisted on the fact that the causes of war could be connected with reasons of economic necessity and opportunity. They drew attention to the presence of dominant, or at least influential, financial groups, who persuaded the people that warlike intervention was necessary and then to stress how disequilibria of economic factors, the cyclical nature of the economy and transitions towards economies of a different kind could all cause wars to break out.

Edwin Cannan, Professor of Economic Policy at the University of London, in his comments on the book by Hirst entitled *The Political Economy of War*, written at the time of the First World War (1915), stated instead that the economic policy of war was a contradiction in terms, because the economic discipline presupposed and analysed an ordered situation, within which operators cooperated peacefully, while war was an active manifestation of anarchy. It was obvious to Cannan that no top rank economist before Hirst had proposed to analyse the situation of warfare and that territory was in fact extraneous to the economy. If no one, for their own good reasons, had thought of studying the economic policy of the civil war, then too studying the economic policy of a modern war between states, those who erroneously held that they had different economic interests did not appear to maintain a coherent idea (Cannan 1927, pp. 49–50). The review published in the "Economic Journal" was later republished in a hefty volume edited in 1927 in which Cannan collected together dozens of articles against war that he had written immediately before, during and after the First World War, where he has analysed various problems of a strictly economic character connected to the state of war.

The contradiction was stressed by Lionel Robbins (1932, p. 7), who attributed the statement that war was irrelevant in the study of economics to the fact that Cannan thought that economics was occupied with the causes of material well-being, while, since war was not a cause of well-being, it could not be an object of economics. Robbins held instead that the task of economics was to investigate the relation between the scarcity of means to satisfy needs that were potentially infinite and thought that Cannan's reserves on war could be considered as the expression of a moral judgement. However, Cannan and his writings more than any other economist had contributed to clarify not only the problems an organized collectivity found itself dealing with in supporting the war effort, but also what could be the best techniques to adopt in obtaining victory in a modern war. In general, in Robbins' opinion, it would be very difficult for those organizing a war to do so without economic knowledge, so the economic policy of war was quite correctly an object of economic analysis, since in wartime a scarcity of means prevailed even more so than its peacetime.

The historian Clive Trebilcock, who dealt with military spin-off, instead stigmatized the approach of pacifists who, by morally condemning war and the economy that supported it, had hindered scientific analysis free from prejudices of an ethical nature.

## The Peaceful World of Economics I

The disinterest of economists in the themes of warlike conflict is particularly evident in university-level teaching of economics, generally undertaken by professors who take marginalist doctrine as their reference point. Neither in the past nor today has war been considered of interest in university textbooks.

Besides, economic history texts, especially the more recent ones, have dedicated little or no space to the war economy, considering it more or less as an anomalous situation not worth taking up because it would not offer anything useful for developing interpretive ideas of a general character.

The war economy was altogether particular, run by the "socialist" state (Mises's definition) and orientated above all to furnish military supplies and obtain victory. It appeared as a break, as a cut-off, which bore no interest for the study of economic laws. While the silence in university courses in economic history and economics in the 1960s and 1970s received some justification, nowadays, war is not even mentioned, even though it still has a major influence on the economies of many countries.

Similarly, the texts for management, company finance and company and industrial accountancy courses do not offer any particular idea about the companies that produce for war, evidently considering them as not unlike those who produce for the civil economy, in their management organization, independently of the fact that they may be state controlled or operate under a semi-monopolistic and/or monopsonic regime.

As far as militarism and the arms race were concerned, along with their influences on the economy are concerned, it fell to Michael Mann (1980, pp. 27–28) to complain about the post-war silence of both liberal and Marxist economists. In particular, he analysed dozens of books and hundreds of articles written in the 1960s and 1970s by authors of the Marxist tradition that dealt with the capitalist state. Mann showed that almost none of them contained a single word on what was one of the principal activities of many capitalist states: "preparing for, or conducting war".

The same result can be obtained from the collection published by Bob Jessop in the "Cambridge Journal of Economics" in 1977 on theories related to the capitalist state and its functions.

As far as the history of economic thought is concerned, Edmund Silberner published a book in Paris in 1939 on war in economic thought from the sixteenth to the eighteenth century, extending this to the nineteenth century in 1957. In it, he complained that apart from his own studies, no *exposé doctrinal* had been dedicated, in France or elsewhere, to an examination of the relationships between wars and political economy. It does not seem that there have been subsequent studies of a general character, except for two very brief summaries, both in French: a "Que sais-je?" series book *La Guerre* by Gaston Bouthoul published in

1959 and the previously cited thick volume by Armelle le Bras-Chopard (1994) concerning theories and ideologies related to war, which includes a brief chapter on economic theories.

In 1971, Seymour Melman, an engineer by training and Professor of Industrial Economics at Colombia University, published a collection of essays concerning the American military economy, the particular characteristics of war industries, their impact on the economy overall and the problem of reconversion of military production to civil production. He stressed in the preface how reflections, concepts and data that were presented on that occasion were not taken into consideration in economics texts. Just as well, studies in industrial management did not offer analyses of the administrative characteristics of military-oriented companies. More generally, he recalled that war and its massive economic implications on both micro- and macro-levels were systematically ignored by economists.

Melman returned to the question of the "peaceful world of economics I" in 1974 (Melman 1974, p. 127), to show how university textbooks for first year economics students mentioned neither the war economy nor the existence of military industry corporations. His observation still holds perfectly true for today's reality. The texts Melman examined in 1974, those in widest use at the time, did not deal in any way with the problem or at most conceded one to three paragraphs to the argument. In particular, none of the textbooks stressed the unproductive nature of economic growth that derived from war investments.

Paul Samuelson provided some space for war in his well-known textbook *Economics* (1948), which had then reached its eighth edition. He criticized the general optimism derived above all from the studies made by Clive Trebilcock on the spin-off technology that cascaded down from the military sector to be adopted beneficially in civil industry. Samuelson also spoke of the inflationary role of expenditure on the war underway in Vietnam, but even he did not make any distinction between productive and unproductive economic growth, thereby underestimating the damage of the war economy.

The peaceful world of economics I, according to Melman, appeared then as "an orderly civilian world, making and exchanging goods and services. In this world, individual consumers and private profit-making firms dominate the field, although there is a 'public sector' and government 'regulates' industry and the levels of economic activity. Military industry, by implications, is one industry among others, and is not differentiated in quality, in terms of control, or in effects on the rest of the economy. The corporation serving the Pentagon is not distinct from other corporations" (Melman 1974, p. 147).

None of the texts provided data on the worrying growth of American state enterprise in the military armaments sector. No one asked if this state capitalism, in its form of "Pentagon Capitalism", was or was not becoming the dominant force in industrial capitalism in the United States and the effects that could derive from this domination. No one faced up to the problem of a new type of industrial enterprise controlled by the state, which, unlike private corporations, did not aim at reducing costs in order to maximize profits. Productivity was obviously recognized as a fundamental aspect in economic growth, but the damage caused by the military appropriation of capital and human resources was not correlated with progress in productivity. For the world of the economic experts in the President's council, the war economy simply did not exist (Melman 1970, p. 151).

The tools required in evaluating the operations of military industries and their management were not provided to the students of the day, as is still the case now, not just for American students, who as a result were and are kept far from making critical evaluations (Melman 1970, p. 151).

Contemporary neoclassical economists do not ask about the relations between the economy and war, either historically or immutably. More simply, they do not ask why war takes place and persists in the capitalist world in a generally freetrading global economy. It is students of other disciplines or pacifist militants who operate outside the university environment who write more to ask what are the advantages of disarmament or rearmament or what are the economic causes and effects of wars.

Nevertheless, recent years have seen the appearance of university courses and textbooks dedicated to peace economics, understood as a branch of economic science, which investigates latent or open conflicts between states, organizations and social groups. The research methods used are the instruments of rational choice and contemporary economic analysis. According to Caruso it is not limited just to the study of conflicts, but instead presents an intrinsically normative nature [...] therefore peace economics enters with full rights in the modern economics of institutions, thanks to the fact that institutions, interpreted as 'the rules of the game' that discipline potential conflicts between rational actors, constitute, in the final analysis, the drivers in long-term of the development of society. (Caruso 2017, pp. 17–18).

## Notes

1. See paragraph VII. 3. below.
2. Christian Schmidt, in his book entitled *Penser la Guerre, penser l'économie* (1991, p. 24), intended to analyze the thought of economists who had made an important contribution to the conceptualization of their discipline and who, at the same time, had integrated war into economic thought. He found himself with a list of just two names: Smith and Edgeworth.
3. Robert Aumann participated in the Econometric Research Program, established by Morgenstern in the early 1960s at the University of Princeton. Years later, he spoke glowingly of Morgenstern who he recalled at a conference on the development of Game theory organized within the programme. One of the speakers at the conference was Henry Kissinger. According to Aumann, Game theory influenced Kennedy's behaviour during the Cuban missile crisis (2008b, pp. 27–28).
4. Richardson's manuscripts from 1947 to 1951 were published posthumously in 1960 in two volumes under the titles *Arms and Insecurity* and *Statistics of Deadly Quarrels.*
5. A group of Israeli intellectuals wrote a petition to the Swedish Academy of Science asking for the Prize awarded to two "warmongerers" to be withdrawn and received thousands of signatures. Schelling was accused of having used his theories to inspire the war conduct of Americans in Vietnam and Aumann of having used his analyses to justify the Israeli occupation of Palestine and of having condemned the decision of Ariel Sharon to withdraw from the Gaza Strip as it was influenced "by a dangerous race to make peace". The Academy replied to sustain that its decisions were based exclusively on the scientific quality of the contributions, without considering their effects. News of the event was published

in various periodicals and is available on line. See especially the article *Calls Grow for Withdrawal of Nobel Prize* published in "The Guardian" on December 15 2005.

6. Donald Sassoon recalled that the theory of rational choice goes back to Bentham, who nevertheless thought that war was madness (Sassoon 2011).
7. Joan Robinson thought on the contrary that "When an idea has once been started it must be pursued without regard to consequences, and once a new weapon or means of attack has been perfected it is extremely difficult to prevent it being added to the stock of means of destruction. The clearest case of this that we have seen so far in the atomic sphere is one of the earliest – the bombing of Nagasaki" (Robinson 1982, p. 263).
8. Many people had opposed the idea that deterrence could avoid war, above all nuclear war, in the previous period. In particular, apart from Melman we quoted above (see page ...), Joan Robinson had stated in the Tanner Lectures back in 1982: "It is clear that for a nation that has an enemy, it is necessary to arm, but it is also true that if a nation has arms it is necessary to have an enemy. To justify armaments, fear and tension have been kept up and each side makes use of the others as a bogey" (Robinson 1982, p. 263).
9. Thus, Schelling arrives, by using mathematics, at the conclusion that Norman Angell reached in 1914, without any theoretical passages (cfr. Le Bras-Chopard 1994, pp. 110–111).
10. Schmidt recalls in particular the seminar organized by the International Association of Economic Sciences (AISE) in Paris in 1982 on the theme of "Military expenditure, growth and economic fluctuations"; the session dedicated to the economic impact of the armaments industry at the AISE Congress in Madrid in 1983 and the creation in France and Sweden of associations of defence economists.
11. See Chapter 1, paragraph 2 on this argument.

## References

Aron, Raymond. 1970. *Paix et Guerre entre les Nations.* Paris: Calmann-Lévy.

Aumann, Robert John Israel. 2008a. "War and Peace." *Prize Lectures in Economics Sciences.* Stockholm: World Scientific Publishing Company.

———. 2008b. *Razionalità, Cooperazione, Conflitto. Intervista sulla Teoria dei Giochi.* Brescia: Morcelliana.

Bernholz, Peter. 1985. *The International Game of Power.* Berlin: Mouton Publishers.

Boulding, Kenneth E. 1962. *Conflict and Defense: A General Theory.* New York: Harper & Brothers.

Braithwaite, Richard, B. 1955. *Theory of Games as a Tool for the Moral Philosopher.* Cambridge: Cambridge University Press.

Cannan, Edwin. 1927. *An Economist's Protest.* London: P. S. Kinf & Son Ltd.

Caruso, Raul. 2017. *Economia della pace.* Bologna: Il Mulino.

Cournot, Augustin. 1938. *Recherches sur les Principes Mathématiques de la Théorie des Richesses.* Paris: Librairies des Sciences Politiques et Sociales Marcel Rivière & Cie.

Cramer, Christopher. 2002. "Homo Economicus Goes to War: Methodological Individualism, Rational Choice and the Political Economy of War." *World Development* 30: 1845–1864.

De Finetti, Bruno. 1963. "La Teoria dei Giochi". *Civiltà delle Macchine* 4: 61–66.

Edgeworth, Francis Ysidro. 1967 [1881]. *Mathematical Psychics.* New York: A. M. Kelley.

Enthoven, Alain, and Smith, Wayne K. 1971. *How Much is Enough? Shaping the Defense Program, 1961–1969.* New York: Harper and Row.

Fréchet, Maurice. 1953. "Commentary on the Three Notes of Emile Borel." *Econometrica* 21 (1): 118.

Hart, Sergiu. 2005. "An Interview with Robert Aumann." *Macroeconomic Dynamics* 9 (5): 683–740.

Hirshleifer, Jack. 1987. "The Economic Approach to Conflict." In *Economic Imperialism: The Economic Method Applied Outside the Field of Economics,* edited by Gerard Radnitzky and Peter Bernholz. New York: Paragon House Publishers.

———. 2001. *The Dark Side of the Force: Economic Foundations of Conflict Theory.* Cambridge: Cambridge University Press.

Hitch, Charles J., and Roland N. McKean. 1960. *The Economics of Defense in the Nuclear Age*. Cambridge: Harvard University Press.

Kennedy, Gavin. 1975. *Economics of Defence*. London: Faber and Faber.

Le Bras-Chopard, Armelle. 1994. *La Guerre. Théories et Idéologies*. Paris: Montchrestien.

Luttwak, Edward. 1986. *On the Meaning of Victory. Essays on Strategy*. New York: Simon & Schuster.

Mann, Michael. 1980. "State and Society 1130–1815: An Analysis of English State Finances." In *Political Power and Social Theory*, vol. 1, edited by Maurice Zeitlin. Connecticut: Jai Press.

Melman, Seymour. 1970. *Pentagon Capitalism: The Political Economy of War*. New York: McGraw-Hill.

———. 1971. *The War Economy of the United States. Readings on Military Industry and Economy*. New York: St. Martin's Press.

———. 1974. *The Permanent War Economy: American Capitalism in Decline*. New York: Simon & Shuster.

Milward, Alan Steele. 1977. *War, Economy and Society. 1939–1945*. Berkeley and Los Angeles: University of California Press.

Morgenstern, Oskar. 1959. *The Question of National Defense*. New York: Vintage.

———. 1960. *The Theory of Games*. London: Routledge.

Neumann, John von. 1928. "Zur Theorie der Gesellschafsspiele". *Mathematische Annalen* 100: 295–320.

———. 1944. *Theory of Games and Economic Behavior*. Princeton: University Press.

———. 1953, January. "Communication on the Borel Notes". *Econometrica* 21 (1): 124–127.

Pizzorno, Alessandro. 1986. "Some Other Kind of Otherness: A Critique of 'Rational Choice' Theorie." In *Development, Democracy, and the Art on Trespassing: Essays in Honor of Albert O Hirschman*, edited by Alejandro Foxley, Michael S. McPherson, Guillermo O'Donnell. Notre Dame, Indiana: University of Notre Dame Press.

Quade, Edward S. 1964. *Analysis of Military Decisions*. Chicago: Rand McNally & Company.

Rapoport, Anatol. 1961. *Fights, Games and Debates*. Toronto: University of Michigan Press.

———. 1962. "The Use and Misuse of Game Theory." *Scientific American* December, pp. 108–118.

———. 1964. *Strategy and Conscience*. New York: Harper and Row.

Richardson, Lewis Fry. 1960a. *Arms and Insecurity.* California: The Boxwood Pacific Grove.
———. 1960b. *Statistics of Deadly Quarrels.* California: The Boxwood Pacific Grove.
Robbins, Lionel Charles. 1932. *An Essay on the Nature and Significance of Economic Science.* London: Macmillan.
———. 1939. *The Economic Causes of War.* London: Jonathan Cape.
Robinson, Joan. 1982. "The Arm Race." In *The Tanner Lectures on Human Values.* Salt Lake City: University of Utah Press.
Samuelson, Paul Anthony. 1948. *Economics.* New York-London: McGraw-Hill Book Company.
Sassoon, Donald. 2011. "Si inizia bene, il finale è incerto" *Il Sole 24 ore.*
Schelling, Thomas C. 1980. *The Strategy of Conflict.* Cambridge, MA: Harvard University. First Edition, 1960.
Schmidt, Christian. 1983. *Conséquences Économiques et Sociales de la Course aux Armements.* Paris: Economica.
———. 1991. *Penser la guerre, penser l'économie.* Paris: Édition Odile Jacob.
Selten, Reinhard Justus Reginald. 1978. "The Chain-Store Paradox." *Theory and Decision* 9: 127–159.
Sen, Amartya. 1986. "Rationality, Interest, and Identity." In *Development, Democracy, and the Art on Trespassing: Essays in Honor of Albert O Hirschman,* edited by Alejandro Foxley, Michael S. McPherson, Guillermo O'Donnell. Notre Dame, Indiana: University of Notre Dame Press.
Shubik, Martin. 1959. *Strategy and Market Structure (Competition, Oligopoly, and the Theory of Games).* New York: Wiley.
Silberner, Edmund. 1939. *La guerre dans la pensée économique du XVIe au XVIIe siècle.* Paris: Librairie du Recueil Sirey.
Singer, David, and Small, Melvin. 1972. *The Wages of War, 1816–1965: A Statistical Handbook.* New York: Wiley.
Smith, Adam. 2007 [1776]. *An Inquiry into the Nature and Causes of the Wealth of Nations.* Hampshire: Harriman House Ltd.
Sorokin, Pitrim Alexandrovich. 1937–1941. *Social and Cultural Dynamics.* New York: American Book Company.
Staley, Eugene. 1935. *War and Private Investors.* New York: Howard Fertig.
Tullock, Gordon. 1974. *The Social Dilemma: The Economics of War and Revolution.* Blacksburg: University Publications.
Usher, Dan. 1992. *The Welfare Economics of Markets, Voting and Predations.* Manchester: Manchester University Press.

Veblen, Thorstein. 1917. *An Inquiry into the Nature of Peace and the Terms of its Perpetuation*. London: Macmillan.

Wright, Quincy. [1942] 1965. *A Study of War*. Chicago: University of Chicago Press.

# 7

# The Benefits of War and the Armaments Industry

## War and Prosperity

Some economists have repeatedly looked into the "prosperity" that war appears to favour and into economic crises, including social crises, that usually break out at the end of the war. The economic benefits for the community, allegedly due to war, concern the fall in unemployment and the increase in overall demand, or the destruction of excess production that otherwise would lead to crises.

When facing the apocalypse after the Second World War, it no longer seemed to be acceptable to take such considerations into account, apart from in the United States, which had not suffered the effects of the war on home territory. However, the economic effects, both positive and negative, could be evaluated during the rapid rearmament, in peacetime, and above all during the Cold War.

We should also remark that, unlike in the period after the First World War, when there was a period of unstable and difficult reconstruction marked by violent social conflict, long-lasting inflation and the hard road back to restarting international trade; after the Second World War, there


R. Allio, *War in Economic Theories over Time*,
Palgrave Studies in the History of Economic Thought,
https://doi.org/10.1007/978-3-030-39617-6_7

was instead the dawn of a long period of exceptional prosperity characterized by economic "miracles" in some countries.

Alan Milward, who we have already mentioned above, stands out among the economists in the west who examined the economic role of conflict and the policy of rapid rearmament after the Second World War. He held that starting back in the eighteenth century, a European state could make the choice in favour of war that was consciously based on an economic policy plan, and that this possibility still exists today. The reasons behind this choice differed in each case. For example, at the end of the nineteenth century, colonial wars and general rearmament had provided a great stimulus to technological development in many industrial sectors, such as shipbuilding, metallurgy and machine tool production. Nevertheless, only highly economically and technologically developed countries at the time were able to build the new complex armaments and this modified the possibility and the conditions of the war itself (Milward 1977, p. 4).

Motives of economic interest seem to be present in the case of the Second World War too. Milward analysed the causes leading up to the conflict and recalled that, even if the factors that drove Germany and Japan were of a political and ideological nature, both countries were influenced in their decision by the conviction that the war could also have provided an economic gain.

Historians and analysts in Eastern Europe go further in maintaining that the Second World War had become an economic necessity for Germany in maintaining the state monopoly capitalism established by Nazism, which seemed to entail essentially both territorial expansion and the destruction of the communist state (Eichholtz 1969, p. 1; cfr. Milward 1977, p. 11). This reading of events holds that Hitler's warpath was agreed to by the main groups in German private industry that had made plans to take advantage of a war of aggression.

A somewhat similar reading was made in 2006 by the British historian Adam Tooze in his work *The Wages of Destruction* that analysed the German economic situation before, during and immediately after the Second World War, and maintained, among other arguments, that the decision to unleash the war had a far from secondary role in the requirement of the country's economy to seize the natural resources required for industrial expansion in view of the challenge faced by the American economy.

The First World War, with its previously unseen dimensions in the use of men, arms and financial resources, saw a closer connection develop between the military apparatus and the industrial system. Pierre Léon stressed the benefits accruing to industry from the war economy:

> The war economy was crowned with success since the objectives it had laid down, that is the mobilization of productive forces in order to serve the war, were achieved by the losers too. Capitalism attained an overall economic rationality that had previously been lacking by introducing a collective goal as the source leading on to all activities. The state was both the promoter and financier, while the spirit of competition that flourished under the name of patriotism was used to block the conflicting interests for the time being. (Léon 1997, p. 48)

This new economic policy, with the resulting mechanisms, modified the economic system, without, however, questioning its basis. The coercion used to carry through the war economy nevertheless disturbed liberal traditions, damaged the interest of certain groups, above all bankers, and did not lead to social peace. This, however, was the case only because the experience of state intervention introduced "in the heat of action" turned out to be too brutal. Thus, despite the profits certain groups made, in November 1918 the financial sectors' wish for a return to normal, to "business as usual", was none other than a wish to consolidate the fortunes made, which even the beneficiaries themselves considered as all too evidently due to wartime profits and social changes (Léon 1997, p. 48).

The First World War left various inheritances. Some of the industrial sectors, which had accumulated enormous monetary reserves during the war, managed to acquire or take over other "more peaceful" sectors that had been left fallow during the war years, but which could come back into operation with the return of peace. There were the mergers in 1920–1921; the new trusts that emerged in that period were called "war babies" in America. Furthermore, the easy gains that only war industry could guarantee were not forgotten, and the leading role of the state was later appreciated for its functions during crises, such as holding down unemployment, and its structural effects, such as public financing of industry (Léon 1997, p. 48).

Stefan Possony, an American economist and military strategist of Austrian origin[1] was decidedly optimistic about the economic role of the war. He attributed special economic merit to the war that was about to break out. Possony held that the war in various aspects was a great deal, not just for some productive sectors, even if he was worried about problems concerning the availability of raw materials, supplies and the adaptation of machinery to war requirements. For example, the adjustment of production to meet war requirements, "the military economy", could have, from his original point of view, highly positive results in the social field as, first of all, it would cut unemployment and thereby public spending on payments to those who could not find work. Leaving aside the great costs of the army, he saw the war as being able to reduce social misery and cut the state budget. Besides this, the war, as the mercantilists had stressed centuries before, contributed to damping down internal political dissatisfaction and increased the possibility of transforming the economy in general (Possony 1939, pp. 213–214).

Apart from this and other particular interpretations, many economists concur that it is true to say that the wars in the 1800s and 1900s helped along the economy, thanks to the increase in public demand, and offered large profits to the sectors which filled this demand. Excepting energy producers, whose productive techniques do not change between wartime and peacetime, the end of hostilities, in the past, for the sectors that had grown fastest during the war there was the prospect of a deep crisis afterwards, because of the need to downsize and convert production.

More recent wars, characterized by major destructive power, have led to strong demand after hostilities to satisfy the requirements of reconstruction. Therefore, rather than the war itself, the post-war period provides advantages and quite often chaotic and unstable periods drag on due to the lack of clear and conclusive solutions: "We live in a period where some industrial sectors make great profits and no longer get these greater profits from the preparation for war, but instead from the construction and reconstruction of a political body" (Desiderio 2008, p. 52).

So war still seems to be economically profitable in moments of crisis, and perhaps even necessary for the growth production.

The vast majority of economists and economic historians agree that the increase in military budgets during the Second World War was the determining factor in overcoming the crisis in 1929[2] and that the first major conflict after the Second World War, the Korean War, favoured economic growth, not only in West Germany and Japan, recruited as American allies against the threat of communism, but also improved the economies of the entire West.

It is equally common to believe that military expenditure can favour economic growth or recovery. However, Jacques Fontanel and Jean-Pierre Matière, who have studied this problem for a long time, tend to make a distinction and dampen enthusiasm:

> in line with the Keynesian theory, the increase in military expenditure is likely to favour economic progress, but all the same does not seem to be able to generate cumulative effects in economic development. Besides, even if some countries could benefit from a new infrastructure thanks to the military effort, this is not the case in general. (Fontanel-Matière 1985, p. 136)

The two authors presented their research at a conference held by UNESCO in 1982 which aimed to investigate and come to a conclusion about the relationship between national defence expenditure and economic development. Many papers presented at the conference showed that it was impossible to provide a clear response to the question. In fact, if war expenditure obviously has different effects in industrialized countries, developing countries and third world countries, in the latter case, where quite often much is spent on armaments instead of in promoting the living standards of the population, the results of arms imports on technological improvement and industrialization appeared to be modest and in any case differed from country to country.

Joan Robinson, who was strongly against military spending, recalled that underdeveloped countries purchased arms from developed countries and thereby entered into a relationship of dependence that often had negative effects on their countries' growth, on the use of resources and the nature of the political regime, by favouring the establishment of dictatorships. More generally speaking, regarding all countries at any level of development, military spending has a greater inflationary effect than productive investment, because it does increase production either in the short period or in the long period. Investment in the armaments sector increases productive capacity even less if it is used rather than left unused, while investment in basic industry allows for an increase in investment and starts a self-generating spiral of growth (Robinson 1979, pp. 121–126).

Kalecki too was highly critical of the positive function of military spending. In an article published in 1943, in reference to the military spending of fascism, he noted that massive rearmament, instead of resolving the problem of unemployment, led to war:

> The fact that armaments are the backbone of the policy of fascist full employment has a profound influence upon its economic character. Large-scale armaments are inseparable from the expansion of the armed forces and the preparation of plans for a war of conquest. They also induce competitive rearmament in other countries. This causes the main aim of the spending to shift gradually from full employment to securing the maximum effect of rearmament. (Kalecki 1943, vol. 14, p. 327)

Kalecki returned to this statement twenty years later in the preface to a collection of his economic papers published in Poland in 1962 to stress that it is not just in fascist states that armaments play an important role in fighting mass unemployment. In the United States, where this process was stronger, some aspects typical of fascism appeared (Kalecki 1962, pp. 7–8).

Those who maintain that war has a positive role in economic growth through war spending often refer to the function of technological spin-off from the armaments sector on civilian industry. The main, but not the only, interpreter of this idea is Clive Trebilcock.

## Trebilcock and Technological Spin-off of Armaments

Clive Trebilcock is an economic historian who wrote several articles on the role of armaments in the Edwardian period, focusing in particular on the influence of military technology on peace sectors during the First World War (Trebilcock 1969, 1970, 1973, 1974, 1976, 1981). Trebilcock is convinced that considerations regarding that period can be extended to all war periods. He considers the importance in terms of industrial development of the spread of advanced technologies, taking place thanks to the military industry at the end of the 1800s and the beginning of the 1900s, can be compared with that of the spread of the railways in continental Europe between 1850 and 1870 (Trebilcock 1974, p. 258; cfr. Segreto 1997, p. 145).

More generally speaking, Trebilcock complains that rational analysis of the effects of the armaments industry has been frustrated by the persistently bad reputation of the sector, by what he calls the syndrome of dealers in death. He believes that western culture is prevalently oriented towards pacifism and so has not allowed for an objective examination of the economically creative role of armaments production. The sector has been accused a priori of feeding international conflicts in order to expand its market, favouring corruption in sales deals of arms soaked in blood and of setting up cartels to force weak governments to reach agreements to purchase armaments at inflated prices.

The English historian states that he does not want to take sides for or against war, as this is not his objective, but instead proposes a consideration of the military industry, like any other type of industry, and to study arms producers in the same way as other businesses.

The main merit of the arms industry, according to Trebilcock, is its capacity to promote technological progress in peace production too through technological transfer.[3] He also considers the transfer of military technology from more advanced countries to less advanced ones as an important factor in stimulating industrial development in the latter. The international transfer of technology would in fact create a positive link between war production, scientific research, economic development and international relations.

Trebilcock questions the statement that armaments producers push countries into war for the simple fact that conflicts, at least large-scale ones, would not be in their interest. In fact, he believes that arms are better sold as a deterrent to war and not during conflict. Combat can also damage armaments producers, for example, when losing foreign contracts, or due to problems in international payments, or due to the requirement to spread out payment over time, or even due to the lack of raw material supplies, the supervision of production by the state and public control of profits.

Another accusation, which Trebilock wishes to reject, is that which states that during a war armaments producers would be in a position to exploit governments. In this case, the state is the sole arms buyer on the national market and has to deal with few highly specialized and organized producers that can push up prices. Trebilcock notes, nevertheless, that in the late 1800s, the period of his analysis, free competition was rare in the civilian sector too. On the other hand, the state monopsony clearly sees that it is dealing with an oligopolistic offer without this necessarily leading to dishonest offers on the part of the producers. Corruption, which armaments producers are accused of, is not practised to a greater extent in the arms sector than in any other. His analysis of the Edwardian period shows how corruption was not so much linked to a product as to certain markets. In the early 1900s, Japan and China were the foreign markets where corruption was most widespread, be it for the sale of armaments or railway equipment or even loan negotiations.

However, what Trebilcock was most interested in stressing was the particularly useful function of the armaments industry in favouring technological development. Starting with an article published in 1969, then in further articles, he proposed the application of the concept of spin-off, previously used by economists, to economic historians.

Historically speaking, the pressure applied to the armaments industry in order to produce increasingly sophisticated armaments leads it to intensify technological research much more than in the case of civilian production. Consequently, the level of scientific knowledge in the armaments sector is often such as to produce innovations in mechanisms and materials that can be useful not only in armaments production, but also in sectors of the civilian economy. This transfer of technological innovation, defined as the process of spin-off, means that military expenditure, although unproductive in itself, as Keynes had already shown, cannot be considered exclusively as waste. In the case of the First World War, expenditure on research in the sectors of warships, explosives and light arms had positive effects in the technological progress made in metallurgy, shipbuilding, mechanics and chemicals for civilian use. Perhaps the most important effect was due to the fact that technical competence acquired in the standardization of war production could, after the end of the war, be transferred successfully to the automobile sector and that of bicycle and motorcycle production. Consequently, generally speaking, Trebilcock agrees with Henry Rosovsky who had analysed the case of Japan and maintained that military investment had multiplier and accelerating effects, and that its impact on heavy industry was both useful and profitable (Rosovsky 1961, p. 22; Trebilcock 1976, p. 98).

Late developing countries too can, as previously stated, take advantage of war technology. When the gap reached between their own production and that of imports becomes very large, underdeveloped countries are forced to recruit technicians, consultants and specialized personnel from abroad. If the main aim of this intervention is to produce more sophisticated armaments, the technological innovations adopted in armaments production can then spread to the advantage of heavy and mechanical sectors, which thus receive an input of advanced technology, so providing a major contribution to the general development of the country.

On the other hand, critics of war maintain that armaments purchases take a large part of the states' budgets and lead to problems of opportunity costs. Investments in armaments are considered as a sacrifice in terms of alternative civilian consumption, usually social services. Armaments are very expensive and in themselves unproductive; therefore, many observers maintain than any advantage from spin-off that could be derived from them is obtained at exorbitant expense. An identical expenditure on non-military uses would produce more positive results in terms of growth. Trebilcock opposed these considerations with a distinction. In the early 1900s, given the level of international tension, defence costs could not be considered as a weight on the civilian economy of the country, but rather as the price to pay for its very existence. This means that net opportunity costs of armaments should be calculated as a lower figure than that of the military budget. Furthermore, in the early 1900s, there was no alternative to military expenditure (housing, medical research, social security, international aid) which are now included among opportunity costs. Finally, it is doubtful if lower military spending would always mean higher investment in civilian sectors. The reduction of military expenditure can in fact simply mean a reduction in public expenditure. Trebilcock evidently did not view this as a positive outcome.

All Trebilcock's articles conclude by restating that the British economy drew a major advantage from the transfer of technology from the armaments sector to the civilian one, even during the decades leading up to the First World War, just as did underdeveloped or developing countries such as Japan, Russia and Italy, which had by then started or were about to start their industrial take-off. The military production projects underway in their shipyards and arsenals were under the supervision of British technicians and manufacturers and the new armaments industries started up on these occasions were large scale and employed advanced technology of foreign origin. This technology then flowed into non-military sectors. This flow was both rapid and large-scale, given the vast military expenditure met by latecomers and small national powers too.

> The spin-off concept thus provides a place for the defence industries within the schema of technological development and, doing so, helps [...] to remove these industries from their position of pariahs of modern economic development and to locate them within an intelligible relationship with the 'civilian' manufacturing sector. (Trebilcock 1976, p. 104)

It remains to see if spin-off of innovation could take place in a cheaper way in the civilian sector. Trebilcock did not seem to think this was the case in particular in the years immediately before and after the First World War. Even in the 1970s, it was difficult to maintain high levels of research in the civilian sector. Few companies had the possibility, the will or the technical personnel able to carry out research that could lead to the realization of innovations of the same type obtained from the spin-off from the armaments sector.

Trebilcock wrote during the final stages of the war in Vietnam, which was also a period of contraction in American civilian industry, and so the advantages of spin-off appeared to be particularly limited and hard to find. Nevertheless, he maintained that it was just a question of time and the effect of spin-off could be realized rapidly again, when the levels of military technology and civilian technology became less disparate. In this case, armaments demonstrated a technical similarity to products for civilian use so that the technological transfer was easier and quicker. This has happened, for example, in the automobile industry after the First World War. However, when the technological gap between war industry and that of peace is very much greater, more time may be required to allow for the realization of technological transfers. Trebilcock admitted that he was writing in a period when the positive effects of military spin-off were less significant compared with 1914 and the calculation of opportunity costs too showed how military expenditure was less favourable for economic growth compared with the beginning of the century, but he warned historians and students of social phenomenon in general against extending contemporary evaluations to the historical reality of the past. He did this just at the moment when he himself was extending the consideration regarding the Edwardian period to general historical reality.

Studies of particular cases of industry have allowed for the checking of Trebilcock's statements, especially when the data available refer to the historical moment he examined, and in particular to the application of advanced technology derived from war to the automobile industry at the end of the First World War. The research by Patrick Fridenson into the history of the Renault company, for example, has demonstrated the decisive role of the benefits coming from the First World War in its evolution. Nevertheless, in this, as is the case for the majority of studies of companies in wartime too, it is difficult to evaluate the exact dimensions of the advantages that the company got from the war, given the discrepancy between the figures obtained from accountancy documents of various types, such as balance sheets, company financial reports and tax declarations. It is in fact very common to find that, especially in wartime, documents produced for different reasons contain very different series of data. Furthermore, the mechanical industries which produce different products, but not all and not exclusively for war purposes, do not make it clear how much profit is made from military production. Lastly, the great monetary erosion during and after the war has to be taken into account. Nevertheless, in the case of Renault, the increase in production during the war was so great that at the end of the war the company became a public company with 80 million Francs capital and started up a trust. Not taking into account wartime inflation, company turnover increased more than fourfold compared with the pre-war period (Fridenson 1998, p. 89). Apart from the immediate benefits and short-term state demand during the war for trucks and airplane engines, in the mid- to long-term period Renault, like the other companies in the same sector, took advantage of the build-up and modernization of the productive system, which was easily converted to motor car production after the war, leading to mass motorization. The war also helped FIAT a great deal in rising from 30th to 3rd in the table of large Italian companies and allowed it to build the ultramodern Lingotto plant, but to what extent these advantages were due to wartime profits and how much to military spin-off has not been made clear. Similar situations developed in the main European companies in the same sector. René Sedillot, who studied the history of Peugeot, agrees with the fact that if the motorcar used the war, the war used, greatly too, the motorcar. This was very much the case in France, where the automobile rose from being an auxiliary industry to reaching the front rank: "from the taxis to the Marne to the trucks at Verdun, the automobile ensured the arrival of supplies, lifted and towed" (Sedillot 1960, p. 113), replacing horses and meshing with the railways in wartime transport. According to the French Ministry of Trade, during the First World War, the automobile industry overall increased the number of factories, machines and workers by at least fourfold (Ministère du Commerce 1919, t, I, p. 328; Fridenson 1998, p. 119).

Another sector which obtained great advantage from the war was obviously the merchant fleet. Serge Ricard (1986) recalls that wars have always been bountiful for maritime transport whose growth, despite losses during the way, is unremittingly helped by conflict. He provides documentation to back this statement with a data series for the American merchant fleet running from the end of the 1700s up to 1970. In Italy, Ansaldo rose to be the largest industrial company by the end of the First

World War, even though its financial standing was so seriously threatened by such excessive growth and reckless speculation by the colossus created during the war, that it was split up shortly afterwards. A great many other examples are to be found in shipbuilding in the countries participating in the war.

The study of Trebilcock's work on the relationship between scientific research, armaments, economic development and international relations was followed by research carried out in Great Britain, France and Germany. In Italy, apart from a work by Luciano Segreto, (1997) on war industry and industrial development, in 1991, an international conference was held in Fiesole entitled "The Armaments Industry and European Economic Development (1870s–1939)". It was not by chance that the historians took up Trebilcock's ideas and almost unanimously agreed with them as regards the First World War.

The analysis of industrialization processes deriving from the transfer of military technology, and from advanced to less developed counties too, has also been taken up by various researchers, sometimes being driven to its limits. Paul Kennedy, an American military historian, wrote a lengthy book, which was very successful in the United States and widely discussed in Europe too (Kennedy 1988), which reconstructed the rise and decline of great world powers over the last 500 years based on their capacity to sustain the high costs required to maintain military supremacy without weakening the economy. Kennedy paid close attention to the relationships between a country's military technology and its economic growth and, even if he admits that his thought "sounds crudely mercantilist", holds that wealth is generally required to underpin military power "just as military power is usually needed to acquire and protect wealth" (Kennedy 1988, p. XVI). When productive capacity increases, the country is placed in the condition to undertake massive rearmament. If, however, military costs are excessive in relation to the wealth produced, national power is weakened. These two extremes mark the beginning of the rise and the decline of a country respectively.

Kennedy does not mention Trebilcock when he maintains that technical improvements taking place in the military field interact fruitfully with civilian technological progress, start a rising spiral of economic growth and ever greater military efficiency, leading the country concerned to overtake the others.

Baran and Sweezy, who wrote about military expenditure in the Cold War period, in 1966, a few years before the publication of Trebilcock's articles, noted how the new military technology developed in that period had reduced the capacity of military expenditure to stimulate the economy. A considerable part of this expenditure, the part destined for atomic bomb production, missiles, rockets and spy planes, was in fact used on highly specialized research, which produced results that could only with great difficulty be transferred to the civilian sector. "Military mass products", such as artillery, tanks, planes, ships and trucks, which had been decisive in the two world wars, and whose technology was easily transferred from wartime to peacetime production, were much less important (Baran and Sweezy 1966).

Christian Schmidt maintained in the previously mentioned 1983 UN report that civilian fallout of military techniques had been greatly exaggerated because, even if certain sectors of the civilian economy, such as metallurgy, naval construction and transport, owed much to war research, in other cases it was difficult to establish if the technological transfer was from the military to the civilian sector or vice versa (Schmidt 1983, p. 86).

There are also those who think that technological development in the military field, instead of providing beneficial innovations for peaceful uses, could on the contrary sponsor war. This is the thesis maintained by Alain Joxe, among others. He believes that the contemporary world is dominated by a particular militarism which he names "research and development militarism" as well as by armaments production. These characteristics have conditioned the range of macroeconomic and micropolitical factors that have constituted since Hiroshima the "corner stone" of industrial development through the close link between progress in production and progress in destruction (Joxe 1991, p. 17).

When dealing with this particular relationship between development, armaments and technology, some people have totally ignored the technological merits of war. Luciano Segreto remarked on a rather noticeable silence regarding war spin-off. David S. Landes' *The Unbound Prometheus* (1969), a book dedicated to the examination of the relationship between technological transformation and industrial development, which was widely read in Europe in the 1970s, totally omitted any reference to the effects of the First and Second World Wars on civilian industrial technology. Segreto maintains that Landes' silence is due to his "ideological prejudice" regarding war (Segreto 1997, p. 145).

## Cold War and Military Keynesianism

The research, creation and testing of means of destruction during the years of the Cold War reached astronomic levels in the United States and the Soviet Union. This expenditure included both atomic and conventional weapons, missiles, planes and spy ships, and the instruments for use in electronic warfare aimed to protect the effectiveness of the country's own electromagnetic emissions, while hindering those used by the enemy by operations of disturbance (cfr. de Arcangelis 1981, p. 363). Aside from electronics, space research too in that period can for many reasons be considered as military expenditure. These sectors saw the mingling and confusion of scientific, economic and military objectives, above all when the programmes were conducted and developed directly under the control of the armed forces.

In the west, some economists wondered if the huge expenditure on armaments in the second post-war period could be considered as a kind military Keynesianism that could bolster the economy, or, alternatively, was not destined to threaten the viability of the state finances of the two superpowers. As regards the Soviet Union and its satellites, where military expenditure depressed the civilian economy and clearly reduced the consumption of the population, this was denied.

Among the analyses in the west that lay claim to the Marxist interpretation, the sociological one by Ernest Mandel stands out with the reading of the *golden age*, of the great post-war growth from 1945 to 1965, as another development of monopoly capitalism typical of imperialism. He called it "late capitalism" even though he was not satisfied by the term, as it refers to a chronology, while he himself would have rather found a word to express a synthesis.[4] Just as in all preceding capitalist epochs, armaments production during the Second World War remained a fundamental element in the economy. In fact, Mandel spoke of a permanent arms economy.

Mandel added on to the Marxian reproduction scheme of two sectors (means of production and means of consumption) a third one for means of destruction which, unlike the first two, produced commodities that did not enter the reproduction process (i.e. they neither replaced nor added to means of production and the labour power required to produce them). This third sector allowed for the diminution of the tendency of capitalism to cause crises of overproduction, even if it could not resolve them over the long term, above all when technical progress is very rapid. In any case, military demand cannot develop forever. When quoting Moszkowska (1943, pp. 117–118), Mandel concludes that when a certain limit is reached, the unacceptable impoverishment of the population, or an imperialist war would intervene (Mandel 1975, p. 302).

Even though the argument considers the rearmament of the 1950s and 1960s, Mandel recognized that it above all provided a powerful boost to the acceleration of technological innovation and not just the most important response to the problem of excess capital.

The permanent arms economy was connected to both internal and external policy and was linked to the interests of armaments producers and those of the military high command and political groups, which have imperialist tendencies.[5] "The 'permanent arms economy' contributed substantially to the accelerated accumulation of capital in the 'long wave' of 1945–1965, but according to Mandel, it was not the basic determinant of this wave" (Mandel 1975, p. 306).

The economic function of the policy of rearmament also included the protection of American capital investments abroad and the protection of the "free world", "free capital investments" and "free repatriation of profits". Armaments also served to guarantee free access of American monopoly capital to a range of raw materials. Mandel quotes a declaration made in 1957 by the President of the Board of Texaco on this matter according to which, from his point of view, the main task of the American government was to establish a political and financial climate, both at home and abroad, that favoured the investment of capital overseas.

A particular judgement on the economic usefulness of armaments was provided in the 1960s by Michael Kidron, who, like Mandel, stated the Marxists principle that capitalism would lead to overproduction and unemployment if forms were not found to rebalance supply and demand. He too found the solution in war spending, concentrating, however, not so much on armaments production as the effects of the tax burden of the state caused by such expenditure. Capital is deprived of resources that could be used for other investment, if it were not taxed to support armaments expenditure. Consequently, armaments for war hold back the fall in the rate of profit and ward off economic crises. Arms are also luxury goods, as they do not serve in the production of other commodities, and, in theory, their production is a factor which slows, perhaps permanently, the tendency of the rate of profit to fall (Kidron 1968).[6] Armaments are the best form of waste, because it is far more expensive to destroy than produce, and because the dimensions of military expenditure are in the hands of the state.

The American "affluent society" of the time, Kidron states, rested basically on high military expenditure, which could counter overproduction, and offer alternatives to capital exports. Kidron held that, in general, a military economy, "once established", became necessary both as regards the threat posed by the adversary and from a more strictly economic point of view, even though he did not analyse the details of the relations between the Pentagon and private armaments manufacturers.

Kidron too admitted that, all the same, economic stability could not be based permanently on the war economy, and interpreted the arms race in that period between the United States, and the Soviet Union in particular, as a full challenge between the two blocks, an economic war in which each hoped that the enormous costs of the war system would lead to the economic collapse of the enemy.

Kidron was convinced that the agreement on prices of military orders put the states at a disadvantage because they did not have sufficient information on production costs.[7] Kidron maintained this opinion, despite the conviction that military contracts remained a supportable cost for the capitalist economy faced with the advantages such expenditure offered, not only for the rebalancing of the economy, but also for the positive fallout that military research had for innovation in the most varied civilian sectors, ranging from air navigation, calculators, pharmaceutical products to reinforced glass, and finally because some branches of industry were essentially based on military expenditure. In this last case, we can find a leading position for aeronautics, whose products and accessories were sold in 90% of cases to the government (in Great Britain 72%) and above all for military purposes. Other sectors, such as chemicals, electronics, metallurgy and optical instruments largely depended on military orders. Even a gradual move towards a situation of peace and demilitarization would consequently push a large part of American and European industry into a crisis.

Martin Shaw too saw the production of armaments as a possible stabilizing factor in the economy. Military industry furnishes products that are final products that rapidly wear out; luxury goods that cannot be used either as means of production for other goods, or for consumption. Their production does not affect the overall rate of profit (Shaw 1984, pp. 61–62), but, on the other hand, leads to a massive increase in employment.

The thesis that war, or at least considerable military expenditure, are required for a favourable economic performance, is explained in a curious paper that appeared anonymously in the United States in 1967 under the title of *Report from the Iron Mountain on the Possibility and Desirability of Peace*, translated into French and published in Paris as *La paix indésirable? Rapport sur l'utilité des guerres* (1968) with a Preface by Herschel McLandress, a pseudonym for John Kenneth Galbraith, generally considered as the author of the entire work. Galbraith denied that he was the author when the pamphlet was republished in 1984, but no one believed him. The work, presumably a work of phantasy, was presented by the so-called McLandress as the real report produced by a commission of experts employed by the American government to make a secret evaluation of the problems posed for economic and social policies by a possible situation of permanent, or at least long-term, peace and to offer ways to resolve them.[8] The commission's analysis concluded that, from an economic point of view, for modern industrial societies, which produce more than demand can consume, permanent peace was not at all desirable. War was in fact the very basis of the organization on which society stood, because military expenditure was the only safety mechanism with sufficient power to stabilize economic progress. The fact that war is waste is exactly why it can carry out this function (McLandress 1968, p. 93).

Thus, in reality, conflicts of interest between states do not lead to war. The root of the apparent opposition between national interests can be found in the dynamic requirements of a system based on war: requirements that imposed military expenditure and the occasional recourse to armed conflict (Galbraith 1967, p. 165).

War had carried out, and still carries out, the regulatory functions that are essential in the consolidation of contemporary society, such as reducing population growth, maintaining political cohesion and balancing the economy. It favoured scientific and technological research too. From an economic point of view, war was an effective guarantee for the control of the surplus product through a massive waste of resources realized outside the normal circuit of supply and demand. No other substitute for military expenditure, such as investment in welfare or space research (which in any case had many points of contact with war) was so controllable by the state and equally effective in carrying out that function. Permanent peace would have had destabilizing effects that would have required fundamental changes in economic choices. Without war, other bodies would have had to be selected to be able to carry out destructive functions with such effectiveness.

The thesis of undesirable peace to some extent draws its origins from Keynes, according to whom armaments spending was sterile, but useful for the overall economy. While unuseful expenditure had a multiplier effect on production, for McLandress it is just the sterility of expenditure on armaments that makes it useful. In the same way, war is useful as it destroys, and not because it spurs new investment.

Benoit and Boulding produced a series of 15 articles published together in 1963 as the final contribution of the Program of Research on Economic Adjustments to Disarmament, sponsored by the previously cited Center for Research on Conflict Resolution at the University of Michigan, to sustain the opposite thesis. The book considered it vital to produce an alternative strategy to the arms race that was economically viable, in order to face the destructive power of the nuclear threat. The authors therefore analysed the economic impact of a choice of disarmament, understood as the opportunity to free human and financial resources in order to realize real economic growth, not bloated by the unproductive demand for armaments. They imagined a decade of expenditure being diverted continuously from defence to civilian areas of housing construction, urban transport, school and health care, and looked into the consequences for armament producers, and production in general, on employment and the financial structure and on the balance of payments. Technological innovation and research in the military field were considered as generally transferable at an advantage to the civilian sector along with the workforce, plant and capital. Plans for state aid, financed by the savings made by the reduction of armaments expenditure, were envisaged for structures that could only be reconverted with difficulty. The savings would also allow for an increase in aid to poorer countries, and thereby further reduce international tension.

Francis Hirst's work (1911) analysed the production and trade in arms and the particular characteristics of the companies involved in production. He immediately showed that it was quite "natural" that large military production companies maintained close ties with the government, seeing that the demand for armaments was mainly from the state. Production took place in a way that was only partly similar to that of monopolies and consequently economic theories that considered the behaviour of peaceful industries did not hold true for military industries.

Arms producing companies also pose major problems of a political nature, as the demand for armaments is greater in wartime and thus war is the *raison d'être* of these companies, their aspiration. On the other hand, in peacetime, the problem of exports of armaments arises as the exports could finish up in countries that are potential enemies.

Lastly, it should be taken into account that there is widespread corruption in the relation between companies in the sector, military high command and the politicians of the countries served. Hirst (1915) provides a long list of cases regarding the corruption by major European armaments companies during the First World War and in the immediate pre-war period.

Baran and Sweezy also, as mentioned above, investigated rearmament during the Cold War. They did so to show that the anticommunism of the 1960s led Americans to spend enormous amounts irrationally in armaments, which all the same had a limited effect in stimulating the economy, while instead it caused a devastating arms race between the two superpowers. This expenditure far from guaranteed security, but instead reduced the probability that anyone could survive a total war (Baran and Sweezy 1966, p. 180). Military expenditure was for Baran and Sweezy, as was the case too for Michael Kidron, Ernest Mandel and other Marxists, connected to the nature of capitalism and served to "stabilize" economic growth and marked the importance of role of the state in economic life.

Sweezy founded *Monthly Review* in 1949 to analyse "political, economic and social problems linked to the aggressive role of American imperialism and the struggle of peoples around the world against it". The review had a book publisher "Monthly Review Press" which produced "easy to read up to date books that documented and provided information produced by revolutionary organizations and militants ranged in the front line against capitalist oppressors and imperialists". In one of these books, *The Age of Imperialism*, Henry Magdoff started out from the Leninist thesis to examine the problem from a practical point of view by studying American imperialism in the 1960s as it acted concretely as an imperialism which drew its origins from the Monroe Doctrine, and did not disdain traditional methods of invasion and the use of force (the Vietnam War was underway at the time). However, its organization had been updated to be based on zones of influence and blocks, the conquest of world markets and the monopoly of raw materials, and "economic aid" too, holdings that influenced the international market and finance with a global reach (cfr. Perrone 1971, pp. 7–8). Magdoff analysed how these forms of imperialism acted behind the mask. According to him, capitalism was and remained an expansionist system, and imperialism was not a choice of capitalist society, but its very "way of life" because "the urge to dominate is integral to business" (Magdoff 1969, pp. 26, 34).

The American historian George F. Kennan, a career diplomat, starting as a functionary of the Department of State and later as Ambassador in Belgrade, opposed the policy of rapid rearmament. He was a moderate, a liberal who opposed the Vietnam War. He was a leader of the Committee for an East-West Agreement in the 1970s and worked for and defended the SALT agreements during the Carter administration. Kennan continued to hope for disarmament in the early 1980s and wished for "rational communication" between the Soviet and US governments.

He considered the nuclear bomb to be "useless", dangerous and did not work even as a deterrent. In the nuclear field, he thought the advantage appeared to be relative and merely an illusion.

During his speech delivered when accepting the 1981 Albert Einstein Peace Prize, Kennan stated that, according to his way of seeing, the nuclear bomb was the most useless weapon ever invented because it could not be used for any rational goal. Nor did it provide effective defence against itself. It was only something that could be used in a display of arrogance or a moment of panic to commit terrible acts of destruction that nobody with a balanced state of mind would ever like to have on their conscience.[9]

Kennan held that the arms race in general did not provide an answer for defence or economic problems, but rather to an internal logic that came from the uncontrollable urge that had taken hold of the great powers when they had entered into competition to construct the greatest armament apparatus of any kind.

## The Military-Industrial Complex and Pentagon Capitalism

The critical analysis of radical American politologists and economists in the Cold War period were concerned not only about the exponential increase in armaments expenditure, but also about the link forming thanks to this increase between the industrial apparatus and the Defence Department, as well as the impoverishment of welfare policies sacrificed to military expenditure.

The dangerous political influence that could be wielded by the "military industrial complex" that was coming into being in the United States was officially denounced on 17 January 1961 by Eisenhower in his famous farewell speech. On that occasion, Eisenhower, after stressing the vital function in maintaining peace internationally carried out by military institutions and after stating that the massive American arms system had a useful effect of deterrence, called for maintaining a guard against the connection, that was the result for the first time in American history, between an immense military complex and an enormous armaments industry, whose influence was becoming pervasive in the economy, in politics and even on a "spiritual" level. Even while he recognized the requirement for such a kind of development, Eisenhower recalled that it was still important to understand the serious implications, because they put in question the keystone issues of society through the involvement of resources, work and the very style of life itself. On a government level, it was vital to guard carefully against the military-industrial complex gaining open or hidden influence, because the potential for a disastrous growth in "misplaced" power existed and would exist in the future too. The weight of this complex of power should not be allowed to endanger freedom and democratic processes that had been won. Nothing could be taken as given and guaranteed. Only a wary and informed population could check that the correct compromise between the enormous military defence industry and the peaceful methods and objectives of the country had been reached, thus guaranteeing both security and freedom.

The definition coined by Eisenhower was adopted in many considerations made during the period of the Vietnam War and more generally during the Cold War period. Galbraith dedicated a chapter of his 1967 book *The New Industrial State* to the argument and two years later denounced the excessive power of the military in the economy and politics of the United States in the short book *How to control the Military* (Galbraith 1969).

William Domhoff, a Professor of psychology enlarged on the theme of armaments, in two chapters of different books, even though he did not use the definition of military-industrial complex, as he was interested in the classes in power in America. In 1967 in *Who rules America?* (pp. 251–275), he wrote about the military, the CIA, the FBI and the enormous cost of American armaments, while in 1971 in a volume on the American upper class in the 1970s, he entitled a chapter "The CIA and the struggle for minds" in which he maintained that the CIA formed part of the upper class aiming at a permanent war economy based on a negative ideology of the absolute enemy, an ideology which is traditionally used by elites in power to justify any action that may be required to preserve their own position and privileges. Domhoff continued by explaining how the CIA pervasively spread its thought, starting from universities, where it commissioned and financed research, to then investigate the methodology of the propaganda used among students and intellectuals who were offered showcases to gratify their narcissism. He also paid attention to propaganda activity undertaken in the world of work.

Richard J. Barnet, who we have already mentioned, was instead directly occupied with the security and control of armaments and in 1969 published a book entitled *The Economy of Death* which accused "the trillion dollar misunderstanding" for armaments spending from 1946 on, and asked how much civilian spending had been cut to buy those armaments, how many of these armaments really were required for defence purposes and how much was required to mount an effective deterrence. The dominating thesis in the United States at the time stated that the power of deterrence worked only when the armaments available were able to ensure the total destruction of the enemy, but in this specific case it seemed logical to believe that the enemy, the Soviet Union in the Cold War period, was making the same calculation and creating an endless spiral in military spending. Barnet analysed and criticized the functioning of the military-industrial complex in the period, including strategies, personal relations and shared interests among the people involved in decision making and working at a public and private level.

The English historian Edward Palmer Thompson (1980), on the other hand, held that the growth of armaments found their own internal logic, even if the production and maintenance of the armaments system have a corresponding social system. The system of nuclear armaments was described by Thompson as "the exterminist mode of production" and he considered it as characteristic of contemporary industrialized society, not only of the western capitalist system, but also of the Soviet system.

However, this theme has been dwelt on longest by Seymour Melman, who maintained that the American economy in the 1960s had by now become a military form of state capitalism. He denounced the enormous growth of economic power wielded by the American military during the Cold War and the Vietnam War and its invasiveness in various fields of civilian life, starting with the conditioning of university research. He also paid particular attention to the close link being set up between the Pentagon and suppliers of military contracts in the period.

Michal Kalecki, when considering the period of the Vietnam War, wrote that the United States was "so deeply involved in it that the big business which embarked upon large-scale investment in connection with this war will be so desperately insisting on its continuation that it will require quite an upheaval to bring it to an end" (Kalecki 1972, p. 114)

This "complex" had been transformed, starting with the arrival of the Kennedy administration, in a hidden way without any announcement or public discussion, into an institutional organization based in the Department of Defence and formed by the armed forces, the armaments suppliers, the secret services, scientists engaged in universities and defence institutions and spokespeople for the military in Congress. The action undertaken by this organization in no way followed national requirements, but rather followed the needs of an expanding bureaucracy that answered to no one. The Defence Administration, along with the Atomic Energy Commission and NASA constituted a sort of great state company, a central government management which controlled nominally private companies, but which in reality were subaltern to defence. Melman considered it an urgent priority to place this enormous complex under control and make its operation clear, as it had caused waste, inefficiency and scandals and which he called Pentagon Capitalism: a structure that was more powerful than any great private corporation, and in fact constituted state capitalism.

Even before Melman, others had drawn attention to the creation in the United States of a block between military industry and the state by means of the Department of Defence. As previously mentioned, in the *New Industrial State*, published in 1967, Galbraith had envisaged the creation of a link between private and public economic structures due to transfer from one to the other of highly qualified technicians, and at the same time had identified a tendency among major companies, above all military ones, to make contact with the government organization, later going on to merge to some degree with it.

Galbraith's writings were strongly criticized by Magdoff (1969 pp. 8, 9) who thought that the massive military expenditure in the post-Second

World War period had allowed the United States to hold on to its leading role in the western imperialist system. Magdoff accused Galbraith of failing to analyse the economic choices of the United States as leading world economic power within an imperialist system. In particular, he stated that he had overlooked the fact that foreign commercial initiatives of large companies which dominated the American economy were largely aimed at the purchase of raw materials at a low cost and the opening up of markets, a typically imperialist activity and source of international tension, when analysing the relations between American corporations and United States military policy. Another point, linked to the first, regards the fact that even if Galbraith stressed the strategic role of military expenditure for the development and security of corporations, he says nothing about the foreign implications of the activities of these large companies.

In the 1970s, Murray L. Weidenbaum, Professor of economics at Washington University, noted how the close and continuous relationship between the military apparatus and the large companies providing it with armaments was modifying the nature both of the American public sector and a huge area of private industry. The government was taking over functions typical of private entrepreneurship, while private enterprise was taking on the characteristics of government branches or arsenals. This was leading to convergence between the two parties, leading to confusion and reducing the difference between public and private activity in a fundamental sector of the American economy.

Melman agreed with this analysis, but held that what was becoming consolidated was not the convergence between public and private, but rather the supremacy of the Department of Defence over private military companies through its control apparatus. Melman showed up the negative affects too of the Kennedy military Keynesianism based on unproductive and parasitic expenditure such as armaments. This neither led to a renewal of civilian productive structures, nor to the creation or improvement of social services, but instead caused inflation and depressed the living standards of less well off classes, and could hinder the process of capital investment in the civilian sector.

Melman listed the impoverishment of social services (housing shortage, high infantile mortality, high percentage of poor people) and the damage suffered by the civilian industrial and technological apparatus that derived from the concentration of capital and advanced technology in the military industry when he considered the enormous military spending of the United States.

Furthermore, the armaments were for use as deterrents rather than for defence and "deterrence is not defense. Deterrence is not a shield. Deterrence is an experiment in applied psychology. There is no scientific basis from which to forecast the probability of the success or failure of this experiment" (Melman 1970, p. 27).

A similar argument was used by Joan Robinson when she referred to the period of the Cold War:

> the very process of building up destructive power contributed to keeping ideological conflict alive. It is clear that for a nation that has an enemy, it is necessary to arm, but it is also true that if a nation has arms it is necessary to have an enemy. To justify armaments, fear and tension have been kept up and each side make use of the other as a bogy. (Robinson 1982, p. 263)

Americans fell victim to the illusion of unlimited wealth in the mid-1960s because the GDP almost reached $900 billion, leading them to believe that it was possible to have both butter and cannons at the same time, while failing to take into account that a significant part of this wealth "is being used for growth that is parasitic rather than productive" to create "a product that does not enter the market place, is not bought back, and cannot be used for current level of living or for future production" (Melman 1970, p. 30). In reality, the armaments did not even serve for defence, because, as mentioned above, America already had the military potential to destroy the entire world population more than once over, but this in no way made Americans feel safer.

Melman thought Pentagon military capitalism had created a new form of imperialism by adding on a parasitic institutional network at home to the exploitation of resources abroad. He returned again and again in articles to an argument which particularly annoyed him as a university professor: the Pentagon financing of university research and the conditioning of studies this obliged by limiting resources for free and independent research. In 1966–1967, 43% of the work by American scientists was fully or partially financed by the federal government and 46.5% of this finance came from the Pentagon, NASA and the Atomic Energy Commission, with the greatest share going to the Massachusetts Institute of Technology (MIT),[10] followed by John Hopkins and the University of California.

Melman recalled that in 1970 three million people worked for the Department of Defence, with around a further million depending on it for research, development and back up activities, while 3,400,000 were enlisted in the armed forces. There were 22,000 private enterprises with contracts with the Department of Defence. The industrial system of the Department of Defence in 1968 produced $44 billion in goods and services, that is to say more than the combined total sales of General Motors, General Electric, US Steel and Dupont.

The American government spent $1100 billion on defence between 1945 and 1970 and this expenditure had distorted the evaluation of growth in the country and opportunity costs in terms of the reduction of funds available to maintain industries, social services and employment. The result of the military economy, whose product cannot be considered either as consumer goods or capital goods, was around 8–10% of the GDP and was considered as part of it. Military expenditure hindered that launch of an economic development programme for 30 million Americans living in difficult conditions and was higher than what was required to help the economic development of Africa, Asia and Latin America (Melman 1971, pp. 1–8).

The data collected by Melman and his opinions were confirmed and repeated in an article by Arthur F. Burns (1971) who took up some of the considerations made. Burns recalled that the industries that produced for defence, especially aerospace, electronics and communication groups, had become the largest in America and had favoured the economic development of cities such as Los Angeles and San Diego, Seattle and Baltimore, where large companies had gained wonderful technological knowhow, thanks to contracts from the aerospatial and defence sectors. However, even if the defence sector had stimulated economic development in some fields, it had held it back in others. Civilian production companies struggled to keep up with the standards of the military sector to pay high salaries to engineers and technicians, thereby being able to update their production processes. Melman gave examples of this and considered that the decline in technology in many American mechanical, railway and merchant shipping industries was due to syphoning off money and scientific talent by the military sector, which could afford to pay high salaries thanks to the rich contracts signed with the federal defence agencies. Melman investigated the cases of Singer and Underwood, which made profits, but were technologically backward. Underwood in particular bought in the then technology of Olivetti (Melman 1971, pp. 122–132).

Ernest Mandel not only looked at the Pentagon, but also at the arms race in western states during the years of the Cold War, and noted how the great intensification of military spending was the determining element in growing state intervention in the economies of industrialized countries. While military expenditure in the west ranged from 5 to 7% of national income before 1914, during the Cold War this percentage rose to 15–25% with maxima even of 30%. Furthermore, many advanced sectors in technological progress, such as aeronautics, electronics, telecommunications, nuclear energy and shipbuilding, worked essentially to fill state orders (Mandel 1969, pp. 75–76).

The melding of economic and military interest and the militarization of science caused by the construction of atomic and nuclear armaments during the Cold War were criticized by the historian of German origin, Georg W. F. Hallgarten in his alarming historical reconstruction of the arms race from the First World War to the 1960s (1972). Hallgarten noted the previously unseen interpenetration between economic and political power and attributed a preponderant responsibility for unleashing conflict to the pressure exerted by large companies in the armaments sector he called "the motors of the arms race": Krupp, Vickers-Armstrong, Schneider-Creusot and Putilov. This interpenetration was kept up in the following years and linked in with the evolution of science in the armaments sector leading to the apocalyptical vision of Hallgarten leading to the holocaust, if not "the twilight of humanity" (Hallgarten 1972, p. 369).

Nevertheless, Hallgarten set the date of the birth of the American military-industrial complex during the Second World War when the United States became the "arsenal of democracy", taking the leading role both in the military and in the industrial sector compared with the rest of the world. His reflections which were made during the worrying years of the Cold War problems, centred on the anxious and unanswered problems of a possible "end of humanity": the building of atomic armaments, the fusion of economic and military interest and the militarization of science.

## The Soviet Point of View

The economists and politicians of the Soviet Block who analysed the causes of war during the first years of the Cold War did not move further on than the Marxist–Leninist interpretation, merely adapting the communist economic and class struggle concepts to the reality of contemporary international politics, reading the major tension of the period as a product of American imperialism and the interests of western armaments manufacturers. The Soviet Union and its allies had the duty to oppose them. The gains of communism had to be defended against both external and internal enemies.

Individual countries lived in a world divided into two parts and took part or supported directly or indirectly one or other of the two blocks. This hostility was made more dramatic by both parties: who is not with us is against us. In this situation, conflict was the norm, but if this had led to the outbreak of hostilities, the key question would have become "who eliminates whom?"

Soviet thought saw the war as functional to the cause because it would have accelerated the historical process leading to the victory of socialism. The First World War had in fact led to the triumph of socialism in one country, while the Second had caused the victory of socialism in a block of countries. The question of being surrounded by capitalism in the second post-war period found its response in the Stalinist order to "mobilize" in the military, political and economic field. This situation was called "aggressive insecurity" by the American politologist William Zimmerman (cfr. Zimmerman 1971, pp. 40–41).

The East German economic historian Eichholtz (1969), as we have already mentioned, followed the view of Marxism–Leninism that the

Second World War had seen the outlet of German nationalism in Nazi imperialism, which in turn derived from the convergence of economic interest of large private industry and the emerging state monopoly capitalism.

> German imperialism was an imperialism which had been deprived of colonies, an imperialism whose development was limited by the financial burdens stemming from the war and by the limitations and controls, onerous to the monopolies, which the victorious powers had imposed, especially on armaments, finances, etc. On that account extreme nationalism and chauvinism were characteristic of the development of the fascist movement in Germany from the start; once in power fascism maintained from its first day an overweening purposeful imperialistic aggression – which had been obvious for a long time - toward the outside world. With fascism a ruling form of state monopoly capitalism had been created which aimed at overcoming the crisis of capitalism by domestic terror and, externally by dividing the world anew. (Eichholtz 1969, p. 1)

The warlike and expansionist objectives of Nazism aimed to dismantle the Treaty of Versailles and to conquer a new vital space, obtaining the support of private monopolistic groups. According to Eichholtz, one of the first projects of this policy was the invasion of the Soviet Union "to liquidate it and appropriate its immensurable riches for themselves, and to create a European 'economy of large areas' (*Grossraumwirtschaft*), if possible in conjunction with a huge African colonial empire" (Eichholtz 1969, p. 63; cfr. Milward 1977, p. 12).

With the defeat of Germany and the end to expansionism motivated by the vital space, the problem was now posed regarding US imperialism, expressed in the Truman Doctrine sanctioning the American right to intervene in the internal affairs of other countries in order to protect its own interests. The Soviet Block opposed this with the Zhdanov Doctrine.

Andrei Alexandrovich Zhdanov, at the time Secretary of the Central Committee, expressed the thesis of the irreducible and permanent conflict between American imperialism and Soviet socialism. The military end economic choices of the west (the Bretton Woods Agreement,

NATO, the Marshall Plan) were interpreted as imperialist, colonialist and antidemocratic manifestations. In particular, according to Zhdanov, the United States proposed stabilizing and consolidating its world domination and imperialism to the detriment of the Soviet Union and other communist countries by using the Truman Doctrine. The United States was, furthermore, supported in this choice by the satellites of the UK and France, as well as states controlling colonies, such as Belgium and the Netherlands, and reactionary and antidemocratic regimes, like Turkey and Greece. The coalition of anti-imperialist forces, the Soviet Block, workers' and democratic parties and movements in the west and national liberation movements in colonial countries were required to fight the American plan of subjecting Europe and world imperialism.

Wars had been, before the creation of the USSR, and continued to be the result of imperialism. Soviet politicians repeatedly stressed the function of the communist block as a bulwark against imperialism and war, as well as being a support for ex-colonies and developing countries against western exploitation. The opposition to imperialism had to be exercised in all areas, political, economic, ideological and military, but depended "in many cases on the tendency of economic competition between socialism and capitalism" (Brezhnev 1976, I, p. 355).

There were frequent calls for the future victory of communism over western capitalism in the early post-war period. These hypotheses remained in the following years, but were increasingly meant to be achieved pacifically through economic competition between the two models that coexisted at the time.

The countries in Eastern Europe reached the end of the war in a situation of dramatic poverty. Soviet governments declared time and again that the objective to be realized through economic planning was an annual growth rate high enough to reach and overtake American industrial production in the short term, denying until the coming to power of Khruschev that the strong and constant increase in the heavy industrial sector and the military one was to the detriment of family consumption.[11] Economists were called to contribute to meeting this challenge and consequently were not directly involved in the problems of war or rearmament. Instead, they concentrated on the preparations of economic development plans, data collection and the analysis and commentary on the results achieved in the various productive sectors.[12]

Khruschev's famous speech at the XX Congress of the CPUSSR is explicit on the themes of the confrontation and coexistence of the two systems: even if capitalism had not been able to block the spread of socialism beyond the boundaries of a single country to become a "worldwide" system, on the other hand, capitalism continued to exist and "the contemporary existence of two opposed global economic systems, capitalism and socialism, which developed according to different laws in opposite directions [was] by now a incontrovertible fact" (Khruschev 1956, p. 9).

When attention passed to the future prospects in international politics, the need for peaceful coexistence of the two systems emerged and Khruschev recalled this repeatedly in speeches at party congresses at elsewhere and at conferences.

Khruschev followed Lenin in 1956 when he warned that capitalism would not fall in the near future, because its general tendency to "decay" did not at all mean that progress was excluded, at least temporarily, in some industrial sectors, in some sectors of the bourgeoisie or other countries. While awaiting this fall, western technology had to be adopted and a study made of what was best in the science and techniques of capitalist countries "with the aim of using the conquests of world technical progress in the interests of socialism" (Khruschev 1956, p. 17).

In later years, Khruschev attributed directly to Lenin the statement that peaceful coexistence was vital. In 1959, at the Third Session of the Supreme Soviet, he stated: "Lenin taught that the working class must, both before and after the seizure of power, understand how to use a flexible policy, accepting compromises and agreements when the situation requires and when the interests of the cause require" (Khruschev 1964, p. 117).

The increase in production in capitalist countries was obtained during the Cold War thanks to the militarization of the economy and the arms race, which, according to western economists, allowed the west to overcome crises. However, this was not the case as major armaments expenditure had penalized other productive sectors and hit workers with the constant increase in direct and indirect taxation. This was far from a way to resolve structural economic problems, being instead a way to draw capitalism closer to "the limit in which the action to stimulate a series of temporary factors becomes exhausted" (Khruschev 1956, p. 20).

The policy of a "position of strength" maintained in the west by reactionary and militarist circles suited the monopolists and billionaire armaments manufacturers, while for the whole of humanity, the arms race constituted a major problem. War psychosis in the population led Americans to an absurd build-up of armaments of every kind and the construction of military bases aimed at a pretended "red peril". Anticommunism was used to hide "the aims of power to dominate the world" (Khruschev 1956, p. 28). Whatever reasons could exist for the Soviet Union to unleash a war of aggression? To who's advantage? The socialist state did not need territories or sources of raw materials, or markets to develop, as there were no social groups in it who could be interested in seizing the land of others or in creating areas of investment, and there were no private individuals in the Soviet Union who could become rich through military contracts (Khruschev 1956, p. 33).

The western world presented the Leninists as "partisans of violence", but this was not so, because it was true instead that "there is a Marxist-Leninist thesis which states that wars are inevitable as long as imperialism exists", but this thesis was elaborated when there was a period when imperialism was pervasive, involving the whole world, and the social and political forces that opposed it were too weak and disorganized to prevent imperialists from unleashing wars. In the meanwhile, the situation had changed: the opponents of imperialism had grown in number and were better prepared, composed as they were of social and political forces in the USSR and elsewhere capable of stopping the imperialists from starting armed conflicts, and, in the improbable case of this happening, being able to mount a strong response.

Khruschev, who proposed total disarmament as the way to achieve lasting peace, recalled the western thesis according to which disarmament would cause unemployment, crises and chaos in the American economic and financial system. He attributed the spread of this thesis to the monopolistic armaments manufacturers and recalled that the resources saved in the armaments sector could be better employed when shifted to the health, education and civilian housing sectors, without causing any detriment to employment.

Then Khruschev made a rather heterodox statement by maintaining that "war is not an exclusively economic phenomenon", that imperialism could not be analysed by investing only the economic interests that produced it, because the decision to unleash a war put into play "class relations, political relations and the level of organization and the conscious will of people" (Khruschev 1956, pp. 55–56).

Nevertheless, Lenin's thesis remained valid and so long as capitalism existed "the reactionary forces which represent the capitalist monopolies continue to press for the adventure of conflict and aggression and can try to unleash war" (Khruschev 1956, p. 57). The "liquidation" of the capitalist system therefore remained the key question in the development of society, but the change would not take place through war because "social revolutions cannot be exported, neither at bayonet point nor on missile warheads [...] It is a problem to be solved by peoples linked to the internal development and the objective degree of maturity of each country" (Khruschev 1964, p. 50).[13]

The aggressive character of imperialism had not changed, but wars were no longer "an inevitable fate". They could be avoided, as long as the forces against imperialism were wary and stood ready on every occasion, making common cause.

During his 1959 visit to the United States, Khruschev came to tell the Americans that their grandchildren would be better off living under a communist regime and did so by imagining that the victory of communism would be without war or other shocks, being the outcome of peaceful coexistence between the two systems during the course of which communism would prove its superiority. "The winner will not be the one which produces a greater quantity of arms of extermination and military bases, but the one that offers man a better life, that is, which comes to satisfy better his material and spiritual needs" (Khruschev 1964, p. 124).

A policy of distension started alongside this. The Soviet Union put forward the proposal of controlled total disarmament, which was not accepted, but talks were started to reduce nuclear armaments.

After the dismissal of Khruschev, the first decade of the Brezhnev era saw the maintenance of a policy, generally speaking, of distension.

Nuclear escalation by the Soviet Union recommenced in the mid-1970s, taking advantage of the period of instability in the United States caused by the end of the Vietnam War, the Watergate scandal and the resignation of Nixon. The two major powers started again to spend stratospheric amounts on armaments, both using the same justification of the need for defence and deterrence. The accusation of imperialism was also reciprocal.

Strong economic growth in the Soviet Union starting in the late 1950s alarmed Americans as the greater resources available were used for military purposes. In an article for Congress in 1959, Harry Schwartz wrote:

> The Soviet Union, with a GDP less than half of ours, is today effectively our equal in the military field and ahead of us in the most thrilling contemporary enterprise for the human race, the exploration of space. If the strength and prestige of the Soviets have risen so high in the world, starting from an economic base which is relatively smaller in comparison to our own, we can only observe the situation with trepidation, as it will develop bit by bit as Soviet production increases, and the economic gap between ourselves and the Soviet Union narrows, which is certain to be the case. (Schwartz 1959—author's translation; cfr Boffa 1960, p. 131)[14]

The same year Howard Petersen, a member of the Committee for Economic Development, who was hopeful of the possibility of American economic growth, asked Americans to take more interest in politics and to make a greater effort, warning them that;

> The success of the continuous struggle against communist imperialism will be determined by faith, determination, the spirit of sacrifice, intelligence and the inventiveness we can display. If we fail, it will not be due to an inadequate economic base, except if future changes in relative economic development do not differ greatly from the ones we are now forecasting. (Pertersen 1959—author's translation; cfr Boffa 1960, p. 129)

Years later, Luttwak too considered Russian imperialism and the wish of the Soviet Union to become the leading world military power, an objective replacing the failure of the attempt to become the leading economic power through economic planning and communism. Luttwak placed this phase between the end of the 1960s and the beginning of the 1970s, when America was faced by a period of domestic weakness and loss of status on the international level. It was then the stage of Russian imperialism, called caustically "the highest stage of Soviet optimism" by Luttwak, came into being (Luttwak 1983, p. 25). When this imperialism was analysed in the early 1980s, it did not appear to be necessarily expansionist in terms of territory and did not even aim at war, but just "peace on its own conditions". Luttwak thought that to combat it once and for all that there was no need to think about unleashing armed conflicts, but rather to collaborate with "the progressive forces" working inside the Soviet Union.

A wide range of studies on Soviet military production have been published in the United States and Western Europe, all drawing the opposition of Soviet scientists, who, however always offer little information on the matter. On the other hand, Russian students focused their attention on the American military-industrial system criticized by Eisenhower and on the Western multinational armaments producers and their relationships with American economic and military policy. I.D. Ivanov (1981) and Aleksandr Vladimirovich Buzuev (1985) are two authors in question. The latter questioned the statement of many western economists that multinational enterprises, including armament producers, would have been interested in peace, because the conflictual situation was harmful to their long-term development programmes. On the contrary, according to the Russian economist, multinationals fomented conflicts by interfering in the internal politics of countries late to develop, and Buzuev quotes the cases of Angola, Mozambique and Benin; or alternatively intervened by supporting military coups, as, for example, in Guatemala, Congo, Chile and Bolivia. Furthermore, multinationals defended their interests by calling on imperialist countries to intervene, either by exerting diplomatic pressure, or by cutting off aid, or even with the resort to force. It was no coincidence that world "hot spots" were almost always linked to the struggle of multinationals for market outlets or areas for capital investment or for grabbing raw material sources.

The multinational armament producers then goaded government to join the arms race and to outline new military programmes, as the major crisis in the civilian economy in the 1990s had, on the one hand, led to an increase in the demand for armaments by states, in part to counter the economic downturn, and on the other hand, had directed the multinationals towards the sector of armaments production. Besides, as was noted by western economists too, these enterprises used the mechanism of the American military-industrial complex to enter directly into the state military apparatus. Buzuev quoted the case, among others, of the Radio Corporation of America, which at the time managed the radar detection system in some regions for the American Ministry of Defence.

Not only armament producers came to be interested in rearmament and war. A direct advantage was also found for component manufacturers, special steel producers and those of special electronic devices and yet other industrial manufacturers. The functionaries in the defence department and politicians often created forms of corruption with these industries, leading to abuses and waste of public funds, of which Buzuev produced a lengthy list. The final result was that between 1982 and 1986, the American government planned to spend more than $1.5–1.6 trillion for military needs (Buzuev 1985, p. 166).

On a theoretical level, Buzuev demonstrated that imperialism is essentially based on economic and philosophical concepts according to which wars are indispensible (Buzuev 1985, p. 187). He referred to an article by Michael Intriligator on the matter published in the *Journal of Political Economy* in 1975, where the author holds that unilateral disarmament of a country would lead to an increased probability of war breaking out as another state with a minimal missile armament could attack it (Buzuev 1985, p. 118). The maintenance of peace would be best served by the existence of a balance of power.[15] Theories such as these, according to Buzuev, expressed the point of view of military corporations.

In practice, the American imperialist war strategies were based on a technical-scientific level reached by armaments and on the economic interest of the military-industrial multinationals, which were the instruments to carry through strategic plans. The western arms race increasingly aimed at the qualitative aspect, pushing research along into more advanced scientific and technological research in the military sector, above all nuclear. The military systems of the western countries were integrated, but everything stood under the American military umbrella.

Only France and Great Britain had an independent productive potential for combat aircraft construction and France alone was able to produce medium- and long-range missiles, but these were armaments that were technologically inferior to the American ones. Japanese missiles and combat aircraft were produced under licence from American concerns, or were fitted with American engines, while Japanese warships carried American armaments (Buzuev 1985, p. 194).

Buzuev recalled Khruschev's argument in his speech at the XX Congress of the CPUSSR according to which the constant growth of the military sector would heighten the contradictions in the capitalist system, because it was at a cost to the civilian sectors that lacked the necessary economic resources, thereby slowing the rate of development. Armaments expenditure was the prime cause of a budget deficit in the western states and caused inflation. It was not at all true to state that building up armaments increased employment. In 1982 in the United States, while there was an exponential growth in military expenditure, unemployment reached a record at over 10% of the active population. A markedly higher level of employment could have been achieved by using the money spent on military scientific research and armaments production in the civilian sector instead (Buzuev 1985, p. 204).

Buzuev obviously does not consider the much lower level of living standards of the Russian population compared with that of Americans and says nothing about the rather high percentage of the Soviet GDP dedicated to armaments. The very costly arms race was constantly justified by politicians as vital for the defence against possible American attack. Buznev uses a declaration by United Nations experts according to whom the arms race, stimulated by western military-industrial multinationals

> creates ever increasing obstacles to an atmosphere favourable to the decrease in the role of force in international relations. Furthermore, it also makes it more difficult to form relations between countries by exercising a negative influence on the volume and direction of trade between them, diminishing the importance of collaboration between states and hindering the efforts made to establish a new international economic order on a more equal basis. (United Nations 1978, p. 58; cfr. Buzuev 1985, p. 213).

Buzuev blamed the Reagan administration for wanting to destroy the strategic military balance that had come into being between the Atlantic pact and the Warsaw Pact, and recalled that at the 1982 General Assembly of the United Nations, the Soviet Union had pledged not to be the first to use nuclear weapons and concluded with the hope expressed for peace movements, western trade unions and the workers movement that in every part of the world, including the United States, to demonstrate their opposition the escalation of armaments and to war in general.

## Notes

1. Possony conceived of the Strategic Defense Initiative, more widely known as the Space Shield, after the Second World War. For Possony, see Bientinesi Fabrizio *An 'Austrian' Point of View on Total War. Stefan T. Possony* in *Economists and War. A heterodox perspective.* Edited by Fabrizio Bientinesi and Rosario Patalano. New York: Routledge. 2017, pp. 7–26.
2. For the question of the United States, cfr. especially Samuel Rosemberg. 2003. *American Economic Development since 1945.* New York: Palgrave Macmillan.
3. Keith Hartley (2017, p. 103), on the other hand, maintains that it is not at all simple to transfer technology from the military to the private sector, because research and development in the field of defence are covered by "secrecy, myths and emotions", while Raul Caruso (2017, p. 125) adds that "military institutions tend to slow down the introduction of any innovations made in the military field to the civil market so as to avoid rivals and enemies taking advantage of them".
4. The expression "late capitalism" had already been used in the past by Sombart (Spätkapitalismus) and by Jürgen Habermas.
5. Here Mandel notes the "military-industrial complex" which will be discussed later.
6. The reading by Kidron is criticized by Mandel (1978, pp. 288–289).
7. Kidron (1968, p. 55) quotes in this case a note by the British Public Accounts Committee in 1966, according to which the government found itself at a considerable disadvantage in price negotiations because it did not have the information the suppliers had concerning the costs of previous production (Second Report 1966–1967, HC 158, 5 August 1966, p. 16).

8. The way in which the "Report" was made public was similar to spying in the Cold War, in which some people dared "to think the unthinkable", making some observers believe that a committee of experts really had been formed by the American Government to evaluate the problems that a situation of lasting peace could have caused and that the Report, dry and cynical (or any other similar report) as it may have been, really had been written.
9. The Prize was awarded to him in Washington in May 1981. The text of the speech is published in: George F. Kennan. 1982. *Possiamo coesistere. America e Urss dalla guerra del Vietnam alla crisi polacca.* Roma: Ed. Gianfranco Corsini. pp. 195–202.
10. Robert Aumann met John Nash at the MIT and heard of the Game theory for the first time from him (Aumann 2008, pp. 21–22).
11. Khruschev declared in the Italian edition of his writings on peace (1964, pp. 17–18): "If the Soviet Union was not so powerful, the flames of war would already be spreading across the world. Does the need to maintain the modern level of the defensive potential of the USSR hold back the improvement in the people's living standards? I must be frank and say it does. Missiles and cannon are not butter, milk, meat, bread or even soup. If we did not have to continuously strengthen Soviet armed forces, we would be able to significantly raise the living standards of our people".
12. On this subject, see, for example, the comments of the Gosplan Institute and I. Kotkovsky and T. Koval in Boffa (1960) on American analyses of the Soviet economy.
13. From the speech delivered to the Celebration of Workers of Moscow on 20 October 1960.
14. The Joint Economic Committee of Congress published a collection of papers in 1959 concerning economic competition between the USSR and the USA. The Institute for Economic Research of Gosplan in the USSR commented on these reports in an article in "Kommunist" (no. 1, 1960) with the title "American economists and competition between the USSR and the USA (Concerning the recent collection of papers of the Joint Economic Committee of the US Congress)".
15. Kenneth Waltz had the same idea, but Buzuev does not cite him.

## References


Aumann, Robert John Israel. 2008. *Razionalità, Cooperazione, Conflitto. Intervista sulla Teoria dei Giochi.* Brescia: Morcelliana.

Baran, Paul Alexander, and Sweezy, Paul. 1966. *Monopoly Capital: An Essay on the American Economical Social Order.* New York: Monthly Review Press.

Barnet, Richard Jackson. 1969. *The Economy of Death.* New York: Atheneum.

Benoit, Émile, and Kenneth Ewart Boulding. 1963. *Disarmament and the Economy.* New York, Evanston, and London: Harper & Row Publisher.

Bientinesi, Fabrizio. 2017. "An 'Austrian' Point of View on Total War. Stefan T. Possony." In *Economists and War. A Heterodox Perspective,* edited by Fabrizio Bientinesi and Rosario Patalano, 7–26. New York.

Brezhnev, Leonid Ilyich. 1976. *Questioni della Politica Economica del Pcus.* Roma: Editori Riuniti.

Burns, Arthur Frank. 1971. "The Defense Sector: An Evaluation of its Economic and Social Impact." In *The War Economy of the United States: Readings on Military Industry and Economy,* edited by Seymour Melman, 111–121. New York: St Martin's Press.

Buzuev, Alexandr Vladimirovich. 1985. *Le Multinazionali e il Militarismo.* Mosca: Edizioni Progress.

Caruso, Raul. 2017. *Economia della pace.* Bologna: Il Mulino.

De Arcangelis, Mario. 1981. *La Guerra Elettronica. Dalla Battaglia di Tsushima ai Giorni Nostri.* Milano: Mursia.

Desiderio, Alfonso. 2008. *Guerre del 21° Secolo.* Firenze: Giunti.

Domhoff, William George. 1967. *Who Rules America.* Englewood Cliffs: Prentice-Hall.

———. 1971. *The Higher Circles : The Governing Class in America.* New York: Vintage Books.

Eichholtz, Dietrich. 1969. *Geschichte der deutschen Kriegswirtschaft, 1939–1945.* Berlin: Akademie-Verlag.

Fontanel, Jacques, and Jean-Pierre Matière. 1985. *La Préparation à la Guerre. Défense Nationale et Développement.* Paris: Colloque Émile-Laugier, UNESCO.

Fridenson, Patrick. 1998. *Histoires des Usines Renault. Naissance de la Grande Enterprise 1898–1939.* Paris: Edition du Seuil.

Galbraith, John Kenneth. 1967. *The New Industrial State.* Boston: Houghton Mifflin.

———. 1969. *How to Control the Military*. New York: The New American Library.

Hallgarten, Georg Wolfang Felix. 1972. *Storia della corsa agli armamenti*. Roma: Editori Riuniti.

Hartley, Keith. 2017. *The Economics of Arms*. Newcastle Upon Tyne: Agenda.

Hirst, Francis Wrigley. 1911. "The Policy and Finance of Modern Armaments." In *Statistical Social Paper*.

———. 1915. *The Political Economy of War*. London and Toronto: J. M. Dent & Sons.

Ivanov, Ivan. 1981. *I Monopoli Internazionali nella Politica Estera dell'Imperialismo*. Mosca: Edizioni Progress.

Joxe, Alain. 1991. *Voyage aux Sources de la Guerre*. Paris: Presses Universitaires de France.

Kalecki, Michael. 1943. "Political aspect of full employment." *Political Quarterly*: 14.

———. 1962. *Szkice o funkcjonowaniu wspolczesnego kapitalizmu*. Warszawa: Berlin.

———. 1972. *The Last Phase in the Transformation of Capitalism*. New York and London: Monthly Review Press.

Kennan, George Frost. 1982. *Possiamo Coesistere? America e Urss dalla Guerra del Vietnam alla Crisi Polacca*. Roma: Editori Riuniti.

Kennedy, Paul. 1988. *The Rise and Fall of the Great Power: Economic Change and Military Power from 1500 to 2000*. London: Unwin Hyman.

Kidron, Michael. 1968. *Western Capitalism Since the War*. London: Weidenfeld and Nicolson.

Khruschev, Nikita Sergeyevich. 1956. *La Politica dell'Unione Sovietica. Rapporto al XX Congresso del PCUS*. Roma: Editori Riuniti.

———. 1964. *I Problemi della Pace*. Einaudi: Torino.

Landes, David Saul. 1969. *The Unbound Prometheus*. Cambridge: Cambridge University Press.

Léon, Pierre. 1997. "Guerres et Crises 1914–1947." In *Histoire Économique et Sociale du Monde*. Paris: Armand Colin.

Luttwak, Edward Nicolae. 1983. *The Grand Strategy of the Soviet Union*. London: Weidenfeld and Nicolson.

Magdoff, Harry. 1969. *The Age of Imperialism: The Economics of U.S.* New York: Monthly Review Press.

Mandel, Ernest. 1969. *An Introduction to Marxist Economic Theory*. New York: Merit.

———. 1975. *Late Capitalism*. London: Verso Books.

———. 1978.*Late Capitalism*. London: Verso.
McLandress, Herschel. 1968. *La paix indésirable? Rapport sur l'utilité des guerres. Préface*. Paris: Calmann-Lévy.
Melman, Seymour. 1970. *Pentagon Capitalism: The Political Economy of War*. New York: McGraw-Hill Book Company.
———. 1971. "From Private to Pentagon Capitalism." In *The War Economy of the United States*. New York: St. Martin's Press.
Milward, Alan Steele. 1977. *War, Economy and Society*. Berkeley and Los Angeles: University of California Press.
Ministère du Commerce, Direction des Études Techniques. 1919. Rapport Général sur l'Industrie Française, sa Situation, son Avenir. Paris.
Moszkowska, Natalie. 1943. *Zur Dynamik des Spätkapitalismus*. Zürich, New York: Verlag der Aufbruch.
Petersen, Howard Charles. 1959. "Lo Sviluppo Economico Sovietico e la Politica degli Stati Uniti." In *La Sfida all'America*, edited by Giuseppe Buffa, 119–129. Roma: Editori Riuniti.
Possony, Stefan Thomas. 1939. *L'Économie de la Guerre Totale. Ses Possibilitès ses Limites*. Paris: Librairie de Médicis.
Ricard, Serge. 1986. *Theodore Roosevelt et la Justification de l'Impérialisme*. Aix-en-Provence: Université de Provence.
Robinson, Joan. 1979. *Aspect of Development and Underdevelopment*. Cambridge: Cambridge University Press.
———. 1982. "The Arms Race." *The Tanner Lectures on Huma Value*, 3. Salt Lake City: University of Utah Press.
Rosovsky, Henry. 1961. *Capital Formation in Japan 1868–1940*. New York: Free Press of Glencoe.
Perrone, Nicola. 1971. "Prefaction." In *L'Età dell'Imperialismo*, edited by Henry Magdoff. Bari: Dedalo Libri.
Schmidt, Christian. 1983. *Conséquences Economiques Et Sociales De La Course Aux Armements*. Paris: Economica.
Schwartz, Harry. 1959. *Comparison of the United States and Soviet Economies. Panel Submitted by Panelists Appearing Before the Subcommittee on Economic Statistics*, Joint Economic Committee, Congress of the United States. Washington: Government Printing Office.
Sedillot, René. 1960. *Peugeot. De la Crinoline à la 404*. Paris: Plon.
Segreto, Luciano. 1997. *Marte e Mercurio. Industria Bellica e Sviluppo Economico in Italia 1861–1940*. Milano: Franco Angeli.

Shaw, Martin. 1984. "War, Imperialism and the State System: a Critique of Orthodox Marxism for the 1980s." In *War, State and Society*. London: Macmillan.

Thompson, Edward Palmer. 1980, Spring. "Notes on Exterminism, the Last Stage of Civilisation." *New Left Review* 1 (121).

Tooze, Adam. 2006. *The Wages of Destruction: The Making and Breaking of the Nazi Economy*. London: Allen Lane.

Trebilcock, Clive. 1969. "Spin-Off" in British Economic History: Armaments and Industry, 1760–1914. *The Economic History Review* 22 (3): 474.

Trebilcock, Clive. 1970. "Legends of the British Armaments History, 1890–1914: A Revision." *Journal on Contemporary History* 5.

Trebilcock, Clive. 1973. "British Armaments and European Industrialization." *Economic History Review* 26: 254–272.

Trebilcock, Clive. 1974. "Radicalism and the Armaments Trust." In *Edwardian Radicalism, 1900–1914: Some Aspects of British Capitalism*, edited by Andrew JA Morris. London: Routledge and Kegan Paul.

Trebilcock, Clive. 1976. "The British Armament Industry 1890–1914: False Legend and True Utility." In *War, Economy and the Military Mind*, edited by Geoffrey Best and Andrew Wheatcroft, 89–107. London: Croom Helm.

Trebilcock, Clive. 1981. *The Industrialization of the Continental Power, 1780–1914*. London: Longman Group.

United Nations. 1978. Economic and Social Consequences of the Arms Race and of Military Expenditure. New York.

Zimmerman, William. 1971. "The Soviet Union." In *Conflict in World Politics*, edited by Steven Spiegel and Kenneth Waltz. Cambridge: Winthrop Publishers Inc.

# 8

# Economic Globalization, Realpolitik, New Wars

The fall of the Berlin Wall was followed by further geographical expansion of the sphere of operations of the capitalist economy left unmatched by a similar internationalization of political power, which remained in the hands of the states, also due to the continuing weakness of international institutions. The order established at Yalta was not followed by a *Pax Americana* but rather by the disorganization of nations and the balkanization of many European regions, the proliferation of violence in the third world and the long-lasting difficulty in establishing a new world order (cfr. Jean 1995, p. 5).

Globalization and financial sector domination of the economy contributed to a further distancing in the relation between wealth and national territory. In fact, if the state's political space requires a precise definition within established borders, and must be permanent, economic space, where production, circulation and consumption of goods and services take place, is changeable and increasingly fast moving. We could also notice the immaterial and volatile character of finance. Therefore, while the state continues to protect its territory, the economy tends to open it up and to abolish borders.



R. Allio, *War in Economic Theories over Time*,
Palgrave Studies in the History of Economic Thought,
https://doi.org/10.1007/978-3-030-39617-6_8

The world economic system finds itself operating in a changing situation and in a heterogeneous environment which sees long since developed countries flanked by others experiencing rapid growth, with yet more that remain poor and have even been further impoverished in recent years.

This diversification could be a handicap leading to tension if the disparity between states turns out to be unresolvable, but, on the other hand, could be an economic advantage, if productive factors could be placed in a position to allow for mobilization.

The more powerful states have faced up to the changes still underway and have therefore managed to maintain the traditional realpolitik, while the small or weak states have seen the situation become more difficult than before. As far as defence is concerned, in a bipolar world these states established ties with one or other of the superpowers through the North Atlantic Treaty or the Warsaw Pact. However, starting in the 1990s, they have often seen the break-up of their internal and external power and sometimes have lost the monopoly of violence and have encountered increasing difficulty in maintaining their prerogatives in protecting their territory and population. As a consequence, there have been cases in which the state has handed over what is still considered a prime task, that of defence, to private companies or contractors.[1]

Non-Marxist students of international relations, politologists, geographers and sociologists have all dealt with the emergence of various types of new conflict over recent decades by taking up an explanation of the persistence of wars by political factors, above all the survival of states, their requirement of defence, their will to maintain power and, above all, the anarchy of their relations. The theory of "realism" has come back into vogue, updated by Kenneth Waltz to "neorealism" or "structural realism", while geographers for their part, above all in France and Italy, have again proposed a "geopolitical" explanation for war. Luttwak has added on "geoeconomics", and Paolo Savona and Carlo Jean have counterposed economic geopolitics, these all being disciplines that consider the relationship between state, territory and war. These debates, most frequently taking place at a journalistic and media level rather than in an academic environment, have not attracted many economists.

## Realism, Neorealism and Anarchy in Relations Between States

"Neorealism", often called "structural realism", follows the "realist" current that dates back to Thucydides and Machiavelli, later being influenced by the thought of Hobbes, and considers as its founding fathers the German politologist Hans Morgenthau and the French sociologist Raymond Aron.[2]

Realist thought flourished in the 1950s as an opposition to idealism, which had from the pre-war period on considered favourably policies of international cooperation put forward by authoritative international institutions, but also utopian liberalism, which despite any indications to the contrary continued to be founded on the hope that free trade would lead sooner or later to world peace.

The realists, so named because they largely based their arguments on *Realpolitik*, went back to focusing attention on the state as the main political agent and on its pursuit of national interest, first of all in the search for power, independently of any moral or ethical considerations. Seeing that rationality must indicate the behaviour of states, ethical norms and moral principles as well as the concepts of ideological superiority, which also change over time, should not condition the relations between states which should instead be handled with due care. This allowed for the selection of the correct behaviour to be adopted each time on the basis of the possible consequences resulting from this selection. Peace could be maintained by avoiding ideological confrontations and crusades, while seeking a compromise respecting the interests of each side instead.

Morgenthau challenged idealism in politics by holding that conflicts were the result of the greed for power which typifies mankind. Since human nature remains unchanged over time, the consequence is that the will to dominate is the main cause of conflict both between individuals and between states and that international politics, just like any other form of politics, is always a struggle for power. These basic convictions lead Morgenthau to define six fundamental principles of realism:

- Politics is governed by objective laws derived from an immutable human nature.
- It is the concept of interest, defined in terms of power that allows for the rational understanding of politics (which, however, is also subject to irrational elements).
- Interest, defined in terms of power, changes over time.
- Moral principles cannot be applied abstractly to relations between states, but have to be considered in the light of concrete circumstances in time and place. The political moral judges an action as correct or incorrect on the basis of its political consequences (and not on the basis of abstract ethical considerations); therefore, prudence should be considered as the supreme political virtue.
- The moral aspirations of a state cannot be identified with universal good, but only with that of the pursuit of the interest of the given state, defined in terms of power.
- Even if a person is a pluralist being, the political sphere is the primary sphere of interest among many others. If politics is to be studied objectively and scientifically, it should be distinguished from the start from morals.

Despite its ambiguity and imprecision, Morgenthau's text was widely read by students of international relations for over two decades. Later, it was felt that a more rigorous methodology was required to support the idea that the state was still the most important actor in international politics. This problem also arose in connection with the desire to challenge more fully the optimistic hope that again came to the forefront in the late 1970s concerning the idea that international cooperation could lead to conciliation between states.

Kenneth Waltz's book *Theory of International Politics* published in 1979 offered a more scientific approach and lead to neorealism or structural realism.

Waltz came to work on this volume after a long period of study starting with his highly successful *Man, the State and War,* published in 1959. In this book, Waltz stated that he agreed with Norman Angell in holding that war did not pay, and that if it was useful for someone, it never had paid off for the whole of humanity (Waltz 1959, p. 224). Nevertheless, war broke out repeatedly, while the attempts to eliminate it "however nobly inspired and assiduously pursued, have brought little more than fleeting moments of peace among states" (Waltz 1959, p. 1). Therefore, it was perhaps better to avoid seeking peace immediately and instead see how the incidence of war could be reduced, examining what could be done to have longer periods of peace than in the past.

Waltz started from the supposition that peace can only be achieved by first understanding the causes of wars breaking out. He sought them by analysing the innumerous and contradictory answers to the question "why do wars happen?" and arranged them in three categories: the first referred to motivations deriving from human nature and behaviour, the second to the internal organization of the state and the third to the characteristics of the international system. He called them "images" because each one brought a different relationship with international relations into focus. However, the main element seemed to him to be the "anarchic system", in which present-day states coexist and interact, because even if there can be many reasons for war, linking them to selfish, aggressive or irrational human behaviour patterns, or instead the warlike actions of individual governments, they are always a question of force used in world politics, and so, if we want to understand that those conflicts which occur for the first two causes can lead to war, it is vital to look at the reality of the international situation.

Twenty years of reflection confirmed this conviction, and in 1979, the *Theory of International Politics* established him as the leader in structural neorealism. Neorealism is based on the postulate of the anarchy of the system of international relations, drawing its presuppositions from realism. States, as the holders of political power, seek security (rather than trying to increase their power) through alliances or by aligning themselves with the positions of a dominant power.

Waltz abandoned the discussion on human nature in 1979 and considered instead that the states in the international system were like companies in a national economy with the prime objective of survival. The behaviour of states on an international level was independent of forms of government and political ideologies; it remained the same and was conditioned by the structure of the international system. Seeing that there is no central international authority, each state has to use its own resources for security without there being a division of tasks at an international level. The lack of reciprocal trust and the consequent anarchy limit the ability of states to cooperate because of the insecurity about the real intentions of the counterpart and the fear that cooperation will lead to favouring other states or create discrepancies (Waltz 1979, p. 107).

Even with economic globalization, states, although they are weakened, still have to offer security for citizens and thus at times have to declare war. Since any state can use force at any time, all states must constantly be prepared, either to counter force with force or pay the price of being weak (Waltz 1973, pp. 10–11).

In 2008, Waltz prepared a collection of his many articles on realism, neorealism and the explanations of the origins of war written over the previous decades. Here, he recalled that the periodic return of war was due to the structure of the international system and that theoreticians explained what historians already knew: war is normal. However, he made it clear that the neorealist theory explains every war not simply by examining the structure of the international political system, but also by taking into account the particular characteristics, different situations and interactions between states in each case in point (Waltz 2008, p. 59). Nevertheless, general explanations are also required and a prime cause must always be found. Waltz recalls that any theory would typically maintain that certain factors are more important than others in defining the relations between them. He recognized that, in reality, everything is connected with everything else and that one aspect cannot be separated from another. But theory must still separate one field from the others to be able to deal with it intellectually. Neorealism establishes the autonomy of international politics in defining the structure of the international political system to allow it to construct its theory (Waltz 2008, p. 56). Wars vary in frequency and in other ways too, but the essential question to be answered is: In which way do changes in the international system influence the frequency of war?

Peace is fragile in the anarchic international situation. Lasting peace requires that potentially destabilizing developments solicit a sufficient response from some or all the main actors in the system. The price for poor attention or mistaken calculations in the anarchy in relations between states is often paid for in blood (Waltz 2008, p. 60).

Neorealism accepts the principles assumed by *Realpolitik* when developing its theory of international politics, but views means and ends and causes and effects in a different way compared to the latter. There are also differences between realism and neorealism. One, which is fundamental, is the question of power. Morgenthau thought it was rational for states to seek to accumulate even more power. He therefore saw power as an end above all and, even though he also recognized that it could be a means too and further admitted that at times nations act for considerations other than the desire for power, he still insisted that when states take action, it is not of a political nature.

Neorealism instead sees power as a means, perhaps a useful one, but which could mean running risks if it is insufficient or excessive. A state's weakness could favour the attack by stronger ones. On the other hand, excessive power could lead other states to increase their armaments or to seek alliances to oppose the dominant state. Since power is a potentially useful means, states that realize this seek to use it in an appropriate way: neither too much nor too little. However, in crucial situations, the attention of the state does not insist on power, but on security.

The main reason for the long peace following the Second World War was precisely because the world changed from a multipolar to a bipolar situation. In a bipolar world (which Waltz imagined would last for a long time in 1979), strong but controlled competition is to be expected. Nevertheless, Waltz admits, bipolarity alone cannot explain the long peace between the United States and the Soviet Union. The deep sense of mistrust on both sides made it easy to believe that some of the crises that broke out in the immediate post-war period would have blown up into armed conflict. Why this did not happen must be understood by looking at another important force in pacification, that is the availability of nuclear arms, which are not common arms, but absolute arms, and as such worked well as deterrents (Waltz 2008, pp. 63, 65).[3]

Deterrence is important, but, as is the case more generally speaking for force, for arms too, nuclear or not, there is a limit to the effectiveness of deterrence beyond which more arms do not mean more security for the country which installs them, nor greater danger for its enemies.

States continue to exist in an anarchic order in which self-help is the most widespread way to establish their own security. Chances for peace increase when the states can realize their more important goals without using force, that is, when war becomes less profitable because its costs rise in relation to possible gains (Waltz 2008, p. 63).

Neorealism has been accused of excessive scientific abstraction and having made the theory of international relations inaccessible and of little practical use. Waltz's position has been considered particularly abstract in the realm of the various currents that developed within the theory because, despite its premises, it only takes into consideration the structure of the international system, without taking into account the reality and characteristics of the individual states that compose it and above all showing no interest in the effects of the action of interstate systems, of non-government organizations or globalization (cfr. Hassner and Vaïsse 2003, p. 25; Williams 2007, pp. 1–9).

This has led in recent years to return in some way to consider the outlook of Morgenthau's realism which is less scientifically structured, but better suited to offer suggestions for diplomacy and to those concerned with international conflicts, even though this reading is opposed by researchers who take more into account the effects of multilateralism, the role of international institutions and the effective cooperation between states.

Other researchers, such as Giddens (1985) and Martin Shaw (1984), while opposing the reading made by Marxists, still note the permanence of war in the industrialized world and have linked it to the persistence of conflictual relations between states, or at least to political factors, and not to the capitalist system instead. The position held by Mary Kaldor (1980) seems to be more complex in the area of the study of international relations as it centres above all on the effects of globalization and humanitarian activities.

The analysis made by Martin Shaw tends firstly to separate the global system of states from the concept of global society as a whole. The premise is that war existed before the formation of states (Shaw 1991, p. 9) and that state existed before capitalism did: in fact, it was part of the framework in which capitalism originated and developed. Capitalism then contributed to transforming this framework. Competition between states through warfare, even though the last resort in resolving conflicts, is inherent in the state system and cannot simply be explained by economic competition. Nevertheless, there has been a tendency over time for the two types of competition to merge in parallel with the propensity for every state to integrate its interests with those of the capitalists within its borders. This convergence, as the Marxist analysis foresaw, has led states to take on the interests of national capitals, thereby reinforcing the links between the state and domestic production. Seeing that the state has "its own exclusive means" and above all detains the monopoly of violence, it has the way to impose its own concept of national interest on the capitalists. Consequently, competition between states tends to dominate, if not overwhelm, competition between capitals. This condition appears most clearly during wars, but is present at all times.

Shaw's conception of the system of states (and their relations) has implications for his view of the world economy, which does not entail worldwide commerce or a real internationalization of capital and is not a free trade economy either. The world economy, according to Shaw, has always been and remains fragmented precisely due to the presence of the state system.

Shaw nevertheless maintains that his approach is not an attempt to separate politics from economics or the state from the mode of production, all the more so since a state has the capacity to dominate other states by military or political means only when it can mobilize sufficient economic resources.

Michael Mann also stressed the need to link the relations between capitalism and war and those between state and war. He too saw militarism as being central to contemporary society, but its centrality did not depend on the presence of capitalism or industrialization, which certainly had revolutionized the organization of war and the means to wage it, but had not affected the military tendencies that had existed before capitalism and which derived, and still derives, essentially from the geopolitical aspects of the social structure. In fact, it was just the production of armaments that would lay the foundations for the start of industrial development, and when laid would greatly increase the impact of war whose causes would still have to be sought in the conflicts between states and not productive structures of a capitalist type (Mann 1984, pp. 28–29; cfr. Shaw 1991, pp. 17–18).

Mann studied the historical series of British state expenditure from 1130 to 1815 and showed that it used between 75 and 95% of public expenditure on war or war preparation. The arrival of capitalism therefore had only a minor effect on the use of state expenditure.

Both capitalism and militarism lie at the heart of contemporary society, but they are linked causally. Their association had the effect of greatly increasing the threat of militarism, as regards both the technological aspect and the social aspect. This threat became common to all expanding industrial societies during the twentieth century, not just capitalist ones (Mann 1984, p. 29).

Given this, Mann stressed how war had remained a "rational" and accepted instrument in geopolitical strategy, which states were still responsible for, because the class organization, which should have carried through the proletarian struggle, had not become internationalized, as Marxists had hoped, but had instead remained at a national level. The twentieth century saw the rationality of war weakening as it was veiled by the emergence of a sort of chauvinism that spread through the middle class and then on to the working class (Mann 1988, p. 159).

Mann declared that he was not interested in discussing the then current interpretations of militarism as he based his arguments on the reading which linked wars to the existence of states and politics. Therefore, he did not analyse the positions of radical Americans of the military-industrial complex of the Pentagon, and he did not consider the thesis of militarism being functional in the growth of the capitalist economy, since it meant a large public expenditure on armaments, nor did he evaluate the opinion considering militarism as a way to divert the working class from its exploitation by capital, nor even did he contemplate the thesis that militarism was required to keep the ever more exploited population of the third world under control (Mann 1988, p. 125).

Anthony Giddens held a similar viewpoint when he held that industrialism, capitalism and state military power are institutional forms that interact in contemporary society and he stressed the major persistence of military power in the organization of the state with the considerable function of armaments in industrial development (Giddens 1985, vol. 2, pp. 3–5). Giddens started from this premise, but made contradictory considerations. On the one hand, he held that there is a rational basis to the thesis that maintains that industrial capitalism has a "pacific" nature, also due to states thereby gaining greater control over society and better establishing their boundaries; while, on the other hand, internal pacification liberates resources to be used in wars between states, that have become more destructive following technological developments.

Robert Kaplan (1997) and Mary Kaldor (1999) instead considered the erosion of state authority taking place in recent years in Africa and elsewhere with the consequent loss of the monopoly of violence. Kaldor stressed how it had sometimes become difficult to distinguish between armies and civilians or armies and criminal gangs during wars.

Kaplan did not agree with what Huntington held, did not think that it was possible to return to a state-centric situation and saw the end of the Cold War being followed by "a future anarchy" (Kaplan 1994).

Mary Kaldor instead hoped that it would be possible to offset war and the fragmentation of violence by creating a "cosmopolitan" government, or, to use Richard Falk's term (1995), a "human governance" formed by "political bodies" standing outside territories and based on "an alliance [...] between islands of civility, noted by Kaplan, and transnational institutions". Kaldor saw an embryonic form of cosmopolitan government in transnational non-government organizations fighting violations of human rights, genocide and war crimes (Kaldor 1999, p. 146).

Even if Mary Kaldor still considered that war remained linked to choices made by the state, nevertheless her analysis of the causes of more recent wars regarded concrete reality and did not refer to political factors alone, but also considered more or less directly economic and cultural questions. The results for economic aspects are not always clear. According to Kaldor, the reasons for wars in recent years can be found first of all in the difficulties that economic globalization has created for states, above all weaker ones, by undermining their territorial sovereignty, thereby destabilizing their monopoly of organized violence and their ability to provide a decent living standard for the population. These difficulties can reach the level of threatening the very survival of the state itself and generate tension and disorder that can cause the outbreak of warfare.

The dimensions and the outcome of contemporary wars, however, differ greatly depending on who caused the outbreak of the war: a "dominant imperial state" or the impulse coming from the "regressive" state militarism within a static or falling national capitalism. Kaldor held that in the first case that the intervention would take place in order to encourage general capital accumulation, as was the case, she asserted, for Great Britain until the mid-nineteenth century, then for the United States. In the second case, the intervention would give rise to a "parasitic" type of war.

Kaldor turns Trebilcock's framework around and considers the defence industry as parasitic in any case, including the large modern armaments corporations, seeing the particularly sophisticated type of technological innovation required. Such innovation determines a great increase in costs and the complexity of production compared with a falling utility leading to a "baroque technology" that can only be transferred with difficulty to peaceful sectors (Kaldor 1981).

As far as the relationship between the state, war and the economy is concerned, according to Kaldor, war is an antithetical face of capitalism because it destroys wealth and implies the interruption of production, but on the other hand is a consequence of the capitalist mode of production. Capitalism needs the state to be present in its support, while the state in turn needs war.

Kaldor paid particular attention to the outbreak of "new wars". The question is no longer that of ideological differences or territorial divisions (economic factors are not given), but rather the counterposing of the agents of a cosmopolitan culture, based on the values of inclusion, universalism and multiculturalism, and a policy of identity based on particularism (Kaldor 1999, p. 6). The author hopes, and she considers humanitarian intervention to be a duty, that international public authorities will be enabled to re-enact and maintain legitimacy at a global level while awaiting the constitution of a cosmopolitan government. This type of consideration and expectation is what neorealist economists fight against.

Mary Kaldor sees a reason for conflict in cultural diversity. Samuel Huntington, who shares the idea of the realists who see anarchy as governing the international system, instead imagined that the violence caused would express itself in coming years not at the level of states (or nations, ethnicities or ideologies), but rather as a clash of civilizations. Huntington supported the opportunity of a globalist policy and the primacy of the United States until 1993, while later he became convinced that that American universalism and multiculturalism could have been a danger for world peace. He thought that different civilizations could enter into conflict with the western one, first of all Islamic and Chinese civilization (cfr. Hassner and Vaïsse 2003, p. 63 ff).[4]

## War, Geopolitics and Geoeconomics

"Geopolitics" is a term dating back to the beginning of the nineteen hundreds and was coined by Rudolf Kjellén, a Swedish politologist and Germanophile, to indicate a type of political analysis developed principally in reference to the influence of spatial factors: "it is the science of the state as a geographical organism as it manifests itself in space. The state considered as a country, a territory or, in a greater sense, as an empire" (quoted by Moreau Defarges 1996, p. 69).

Numerous definitions of geopolitics were provided later on: geographical consciousness of the state, the science of power in space, geography of the prince or even voluntarist geography because it sought to define public policies aimed at modifying existing geographical frameworks. More simply speaking, going back to Napoleonic thought, some held that "the politics of a state lies in its geography". This threat became common to all expanding industrial societies during the twentieth century, not just capitalist ones (Mann 1984, p. 29).

Leaving aside the question of definitions, the discipline seems to have been settled with some identity problems since the discussion over its field of action has been going on for some time, above all in the second post-war period. Geographers who have taken it up are, nevertheless, in agreement in maintaining that geopolitics is not a type of political geography, even if the two terms tend to be confused in the English-speaking world.

Karl Haushofer, the leading interwar German political geographer, considered that "political geography observes 'the forms of being of the state' while 'geopolitics is interested in political processes in the past and in the present'" (cfr. Moreau Defarges 1996, p. 72). Carlo Jean instead thought that "political geography concerns the spatial distribution of political phenomena and their influence on geographical factors", while "geopolitics studies the opposite relationship: that is the influence of geographical factors, both physical and human, on the political analyses, choices and actions in relation to those of other political bodies operating in the same territorial area" (Jean 1995, p. 12). According to Moreau Defarges, the French politologist, geopolitics poses questions about the relationship between space (in all senses of the word) and politics, and asks "How do spatial data influence the politician or politics? Furthermore, how does the politician use space?" (Moreau Defarges 1996, p. 7).

Perhaps the definition which best explains the problem of conflict as the domain of geopolitics is that of Pascal Lorot, a French economist and politologist. He considered:

> Geopolitics is a particular methodology which selects, identifies and analyses conflictual phenomena and the offensive and defensive strategies focusing on the possession of territory within the threefold observation of the geographical environment, in both the physical and human sense, of the political arguments of the contenders and the weighty and constant historical tendencies. (Lorot 1997, p. 91)

Since geopolitics analyses the behaviour of the state and poses question about power, it is a theory connected with conflict. As a science, after its development in the first half of the twentieth century, it was put aside during the decades immediately after the Second World War when it had earned a very poor reputation. It was, in fact, considered as a German theory (because its leading exponents were), or even a Nazi one, as it was accused of providing a "scientific" alibi for Hitler's "vital space". This was especially the case following Haushofer's ideas about Central Europe being a natural area of expansion for Germany, or those of his noted predecessor, Friedrich Ratzel (1844–1904), who saw Imperial Germany suffocated by its frontiers and observed that not all Germans lived within them. Furthermore, Germany did not have a colonial empire of the same dimensions of those of France and Great Britain. In fact, according to

Ratzel, for a power to be considered as global, it had to be present in every corner of the world, above all in strategic zones.

The infamous German *Geopolitik*, as amended by its unscrupulous proponents to become géopolitique/geopolitics, was adopted above all in France and Italy after the fall of the Berlin Wall. So, following "regression", or rather "the metamorphosis of the ideological factors", which had conditioned politics over the years of bipolarism (Moreau Defarges 1996, p. 95), there was a return to geography in the attempt to find an explanation for the upsetting of international equilibria, wars and the resulting new period as it emerged. The term has been used a lot and employed with different, often imprecise, meanings to the point that every author dealing with the matter provides a premise to his or her field of research.

In Italy, the tradition of political geographical studies, which from now on will be named geopolitics, goes back to the period of the *Risorgimento* and considered the problems of territory and politics that emerged in the process of unification. Fascism later took up the matter in order to produce colonialist propaganda for "the place in the sun". In recent years, the economist Paolo Savona and the general Carlo Jean, who is also a Professor of Strategic Science, have shown an interest in geopolitics and geoeconomics.

In France, geopolitics developed rapidly after the end of the 1970s and a "left" reading of it was provided by the geographer Yves Lacoste in many articles and the journal "Hérodote", which he founded and edited. Lacoste, an expert in third world problems, did not consider geopolitics as a science and it should not try to formulate laws, but instead it should think of the struggles underway in the area in order to understand what is about to take place and then suggest ways to undertake an effective intervention.[5] In the *Dictionnaire de Géopolitique* (1993), Lacoste considers that present-day use of the term geopolitics means "the rivalry of power over territories and over the people living there". To be more precise, geopolitics can be considered as the situation in which "two or more political actors contend for the control of a given territory. This contention means that the population of the territory, or those who are represented by the actors contending it, have to be involved through the use of mass media" (Lacoste 1993; Bettoni 2009, p. 23), because both the means of communication during the conflict and the reaction of public opinion are far from being uninfluential on the way the war is conducted and its outcome.

According to Lacoste, not only geopolitics but geography too is closely connected with war. So, one of his books, published in 1976, bears the title "*La géographie, ça sert, d'abord à faire la guerre*". Geography can be used to make war because war takes place in the space forming a natural limit to the fighting as it is the territory for drawing up forces and forms the contended objective. Geography is used by those commanding military operations, for those who decide on troop movements and the positioning of armaments, but it also serves for those who prepare for war by constructing lines of defence, establishing lines of communication during the conflict and setting up strong points, in the past, nowadays missile or atomic bases. Lacoste considers that it is just through reflections on war that "geography is done" (Moreau Defarges 1996, p. 97).

If the range of geopolitics is fluid, that of geoeconomics is even more so, seeing that its specific nature is often controversial and contested. The same label of geoeconomics covers studies with very different frameworks, which can often consider the activity of private organizations, above all multinationals and not just the state.

Moreau Defargeswas also the first author to make this distinction by observing the fact that geopolitics studies the relationship between *homo politicus* and space "and, at least at the start, is the product of an epoch running from 1870 to 1945". The end of the last century saw the emergence of *geoeconomics* following in the tracks of geopolitics. Geoeconomics studies "the interactions between *homo economicus* and space: the importance of spatial factors in human production and distribution and the use of space by mankind in carrying out economic activities". Geopolitics considers the operation of the state, while geoeconomics looks further than the action of the state to observe that of industry, banks, producers and consumers too (Moreau Defarges 1996, p. 145).

According to the author, many people in past have engaged in geoeconomics without realizing it. He cites the case of the great French historian Fernand Braudel, above all his books *The Mediterranean and the Mediterranean World in the age of Philip II* (1949) and *Civilization and Capitalism: 15th and 17th centuries* (3 vol. 1967–1979). He also saw geoeconomics emerge and define itself at the end of the twentieth century for various reasons: the *world-economy*[6] had been transformed for the first time into the global economy, there had been a multiplication of interdependencies and economic and non-economic networks with a tremendous acceleration in communications and immaterial transfers which seemed to signal the end of geography because space and time seemed to have disappeared, the Iron Curtain has been ripped aside and "the communist experience" has gone with it, and world space is no longer crossed by fixed lines as new conflicts that are proliferating rarely take place between clearly defined actors (states, blocks). The power of a state, or even its survival, no longer depends so much on its armed forces, but on its capacity to adapt to technical and economic competition, and geoeconomics could suggest the best-suited movements and behaviour patterns.

We should also consider the pervasive and determining presence of "heterogeneous actors". Even if historically speaking the sovereign state has never been the sole actor on an international level, given the activity of the church, the community and non-state institutions, at the end of the twentieth century different factors intervened at the same time to weaken state power. The state control of frontiers has become less effective; large corporations or large productive sectors, such as banking, communications, air transport and vehicle manufacturing, have organized themselves on an international level to follow the dictates of the global dimensions of the market; finance has gained the upper hand over the real economy and has seen collective operators, or even individuals, establishing themselves unhindered at a global level. This has led to "an anarchic economic world, where the dividing lines between what is allowed and what is prohibited, between legal and illegal move and blur continuously". Geopolitics should cast light on these changing and turbulent scenarios where the state, which still maintains particular prerogatives and responsibilities, finds itself threatened and destabilized by actors (companies and individuals) for which theoretically it is the ward and protector (Moreau Defarges 1996, p. 147).

Geoeconomics, which tended to substitute for geopolitics in formulating responses to the questions of the political power of states and the places and ways in which this power is manifested, today no longer seems to attract much interest, while it was "the fashion" at the end of the last century, and Moreau Defarges recalled that no fashion is innocent: on the contrary a fashion expresses something essential in an epoch (1996, p. 153). The term geoeconomics was coined by Luttwak[7] in an article published in 1990 at the end of the Cold War counterposing it to geopolitics in order to intervene in a debate on the difficulties encountered by national states in controlling an increasingly globalized economy and finance whose economic space and market decreasingly matched the territory of the state. Luttwak therefore maintained that military force no longer defined the international importance of a state, but its economic strength, such that future conflicts would be economic not only in their ends but also in their means. Commercial and industrial domination, which today is the goal, was in the past "overshadowed by the more pressing priorities of war and diplomacy, chiefly the quest for security – a sufficient reason for many wars – but also the pursuit of glory or internal political advantage by single rulers or ruling groups, by ambitious individuals or entire castes" (Luttwak 1999, p. 138). In the new era of geoeconomics, however, not only the causes, but the instruments of rivalry too would be strictly economic.

So, for Luttwak, geoeconomics was simply the logic of hostility applied to the rules of commerce. The traditional policy of force used the means of the army, diplomacy, propaganda and the development of armaments and had as its objectives territorial security, expansion, influence over other states and international prestige. Geoeconomics uses public instruments, but private initiative too, be it directly or assisted by the state. The public means of conflict appear as import duties, regulations and quotas or trade barriers, export support finance, economic and technological espionage and counterespionage. Private bodies, helped by the state, can employ policies of dumping, increasing production to win a greater market share and investing in research and development in economically strategic sectors.

All in all, the aforementioned politics can be considered as economic warfare, which apparently is nothing new. This kind of behaviour was already typical with mercantilism, even if "Mercantilism was an economic phenomenon, but its purposes were strictly political, indeed strategic" (Luttwak, 1999, p. 140), which sought to accumulate gold and, wishing to increase the prosperity of the state, led to war and the attempt to damage the enemy. However, today's geoeconomics proposes the achievement of well-being among citizens, thus tending to avoid violence, which mercantilism instead solicited and which now has become rather unprofitable economically, at least for industrialized countries which resolve their conflicts with the instruments of geoeconomics. Combat still takes place in underdeveloped countries, which are unable to carry out warfare beyond their boundaries. Luttwak foresees the possible outbreak of local wars in these countries, and even if not particularly sophisticated arms would be used, they could still be very bloody.

> In the backwaters of world politics, where territorial conflicts persist, war or threats of war provide an ample outlet for hostile sentiments. But when it comes to the central arena of the world politics, where Americans, Europeans and Japanese collaborate and contend, it is chiefly by economic means that adversarial attitudes can now be expressed. (Luttwak 1999, p. 128)

There is, however, a third typology of states, those at an intermediate level of development, present in Southeast Asia, which, according to Luttwak, find themselves at a level similar to European states in the 1800s. International conflicts in these states would become wars between states.

Luttwak, an expert in international strategy and politics and a consultant of the American Ministry of Defence, is a rather prolific writer, frequently appearing on television in various countries, not always coherently. He is convinced in deterrence and has, among other things, supported the decrease in state intervention in the economy, seeing that the impossibility for it to handle or at least channel forces that are now slipping from its control is not at all a negative factor. As a free trader, he appreciates that the state should maintain only its essential traditional functions, including the exercise of violence, even if the end of the Cold War had led to a lessening of the importance of the armed forces in relationships between states.

In Italy, Carlo Jean and Paolo Savona consider Luttwak's approach to be too similar to that of international political economy, or alternatively "too militarized", and have proposed the study of "economic geopolitics", which, differently from Luttwak's geoeconomics, does not just concern "the strategy of competition or cooperation between the state systems, but also the measures of institutional and social engineering required to make their own countries more competitive, with greater solidarity between generations and being better equipped for rapid growth" (Jean 2008, p. 12).

Luttwak's theses have drawn many criticisms. According to Krugman (1991), for example, international competitiveness has not got the degree of importance Luttwak attributes to it. A country's economic fortune is determined by many internal factors. The emphasis Luttwak places on the role of competitiveness would have the aim of gaining internal consensus in the United States for an expansive economic policy. In fact, Luttwak analysed the behaviour of American governments to conclude by maintaining that their political and economic choices can be exported and imposed throughout the world, thereby forcing developing countries to adapt to the social and cultural standards of the West.

Robert Solow (1993), on the other hand, criticized the simplistic application of military-type strategies and techniques to economic reality, as Luttwak does, that is, without even taking into account the different complexity of economic behaviour compared with the military one. States cannot control the economy with the same means as those used to control war activity, for which, we should note, they have a monopoly. If they try to do so with protectionist policies of a mercantilist type, they end up by damaging the whole community. The army and the bureaucracy are constituted as a pyramidal structure, while the economy operates with a network type of organization within which the actors follow their particular objectives. Luttwak does not allow for the possibility that, at least sometimes, trade between nations can be of mutual benefit and, more generally speaking, Solow accuses geoeconomics of truculence.

Lacoste, on the other hand, insists on the fact that territory still maintains its importance in international relations and conflicts. He limits himself to recalling that the Israel–Palestine conflict, with all its dramatic consequences both internal and international, "has at stake a very small territorial area" characterized by particular geographical, geological and above all historical facts (Lacoste, preface by Bettoni 2009, p. 13), when considering those who maintain that geopolitics is on the way out, not only because of economic globalization which leads to integration in a network system, but also because of the emergence of industrial-financial colossuses operating worldwide that are able to condition states too, and also when considering Fukuyama and others who think that history itself is coming to an end thanks to the victory of the West's values and that wars will be limited to those countries remaining "outside history" (which are, however, still the majority). Other students (but few economists among them) like him link the continuity of war to the relationships between states and add that "even if it is true that the majority of states are losing their weight and role, some (few but powerful) remain to continue to act on the world stage in line with their own national interests, that is, with classical geopolitical and geoeconomic schemas" (Chiesa 2000, p. 85).

Stanley Hoffmann takes another view when he considers the effects that globalization in all its forms could have on tension and the possibility of war. Economic globalization is far from being pacific and could create inequality, both between and within states, and might give rise to worry about competition on a global level. Cultural globalization, an effect of the technological revolution and economic globalization, could lead to uniformity through the transmission of cultural products. The reaction would produce "the disenchantment of the world" or "the renaissance of local cultures and languages", as well as attacks on western culture "considered as arrogant and the vector of an ideology composed of lay and revolutionary values used by the United States in its domination" (Hoffmann 2002, p. 61). Lastly, globalized politics, the result of the other two (cultural and economic), could adopt different forms: either American supremacy or that of international institutions.

The problem lies in seeing how the anarchic society composed of states and the open society created by globalization can interact, seeing that "economic life exists on a global level, but our identities remain national" which leads to strong resistance to cultural homogenization (Hoffmann 2002, p. 63), a theme more current than ever. War still exists despite it all, partly due to the fact that the UN and other international institutions have neither the strength nor the capacity to prevent conflicts from breaking out at a regional or international level or from them being resolved with traditional systems of clashes or alliances. The need to accumulate consistent funds for the purchase of armaments leaves poorer states without resources, which should be used to reduce inequality in the population's living standards. The inequality accentuated by globalization feeds conflicts and resentment to the extent that they enrich the few and push aside the majority. So, those who are both impoverished, at least comparatively, and pushed aside may try to seek revenge in terrorism (Hoffmann 2002, p. 63).

## Public Choice and War

Public Choice, or the economic theory of public choices, developed as a field of research in the United States at the end of the Second World War. Generally speaking, it draws its roots from two books by Duncan Black *On the Rationale of Group Decision Making*, published in 1948, and *The Theory of Committees and Elections*, published in 1958, which analysed electoral behaviour, even if the book by Buchanan and Tullock, *The Calculus of Consent*, published in 1962, is more famous and more frequently quoted.

The theory applies economic and mathematical instruments and methods to the study of politics and, more generally, to collective decision-making processes outside the market, thereby extending the behaviour of *homo economicus* to political choices. It studies in particular the economic behaviour of political institutions and conceives of politics as a process of exchange, an agreement or a contract able to determine spontaneous order.

The need for these studies derives from the fact that there was the lack of a theory of economic behaviour of government that could be tied in with that for other economic subjects: the family and the firm. This need was more strongly felt when state intervention in the economy became increasingly important, if not pervasive, both during wartime, for contingent reasons, and later following the introduction of Keynesian policies.

Among the main exponents of the school, we can name James Buchanan, the founder of the journal "Public Choice" and winner of the

1986 Nobel Prize for economics precisely for this kind of approach, Gordon Tullock, Duncan Black, Anthony Downs and Dennis C. Mueller. They recalled Knut Wicksell, whose 1896 book *Finanztheoretische Untersuchungen* [Essays on Financial Theory] dealt with government activity as a political exchange, formulating a principle of utility that links taxes to public expenditure and warns economists against the presumption of being empowered to offer advice to the benevolent despot who seeks to maximize the social welfare function (cfr. Buchanan 1986, p. 23). In fact, Public Choice holds that those who work in the political sphere do not seek the common good, but personal interests, instead, are subject to pressure exercised by interest groups. This opinion had already been expressed by Bastiat and Ferrara.

Another reference point for the theory was the work of de Viti de Marco and more generally the voluntarist orientation of the Italian school of financial science.

One of the more fully developed themes of the school is that of the creation and winning of consent in a democratic regime, as in this case the electors select their own representatives, who, in turn, make the political choices. Public Choice theory hypothesizes political behaviour coherent with the interests of all those involved in the political world: electors, politicians and bureaucrats. Electors vote those candidates who they think can fulfil their interests, while politicians forward proposals that will get them elected and bureaucrats seek to follow their own career paths. So if all act selfishly, the *Calculus of Consent* (Buchanan and Tullock 1962) can be made possible by applying those instruments typical of economic research, such as the maximization of utility, Game theory and Decision theory. The authors make it clear that the book's theme is the political organization of a free society, but the methodology employed is derived essentially from the study of the economic organization of such a society.

The analysis (as Buchanan defines it and as is usually carried out by the school) is positive in the sense that it studies reality (what is), but not infrequently is also applied with normative aims to analyse how politics should function (what ought to be) or to suggest how a system could be improved.

Nevertheless, if politics is to be reformed in a serious way, that starting point should be the fundamental rules underpinning decision-making within society. Therefore, the starting point is the constitution itself, which sets limits to the activity of agents, be they political or economic. Buchanan and Tullock therefore tried to define an "economic theory of political constitutions", that is, a series of economically rational criteria forming the bases for social choices. Buchanan expressed it thus: "It is only when the *homo economicus* postulate about human behavior is *combined* with the politics-as-exchange paradigm than an 'economic theory of politics' emerges from despair" (Buchanan 1986, pp. 25–26).

The economists of Public Choice opposed the expansion of public debt and "deficit democracy" created by Keynesian policies before and after the Second World War by extending their research to the effects on the economy of the state bureaucracy and the irrational choices of voters and to the formation of situations of "rent-seeking" that emerge when there is the contemporary presence of both the market economy and the public economy. In this case, the thesis is developed to mean that government agents and interested private economic entrepreneurs seek to establish forms of privilege, so as to share out the benefits in terms of monopoly income. Situations of "rent-seeking" entail the redistribution of the income that interests politicians to the extent that they can use it in order to gain consensus.

As regards war strictly speaking, Tullock's position is believed to consider public violence just like private violence as a rational phenomenon, which is useful for some, but generally costly for the community because of the losses that it entails, even though it can be analysed with economic parameters.

More generally speaking, these economists refute the characteristic of public good for defence and hold that, in this sector too, a market, even if imperfect, is still preferable to the action of the state which is always marked by inefficiency and bureaucracy.

Making defence subject to competition, as is the case for public goods, would lead to a rationalization of the sector. The offer of war would derive from private enterprises that could offer goods and services to be selected by "consumer-taxpayers", on the basis of price-quality and cost benefits. Defence, which the state nevertheless is bound to offer citizens, following criteria of economic efficiency, could or should be entrusted to private agents who can guarantee greater flexibility, superior professional training of combatants, probably lower costs and certainly a lower death rate to reveal to public opinion.

It may seem to be a provocation, but as some French Marxists have noted, recent history seems to have proven them right: during the most recent wars, states have often "outsourced" combat itself, and we should consider the end to compulsory military conscription and the employment of mercenaries, the noted "contractors", whose presence has been increasingly frequent in the latest conflicts. We should additionally recall the integration dating back to the Vietnam War of private groups in the war effort, such as Lockheed and General Dynamics as regards arms production, Halliburton and Vinnell as regards logistic support, Pacific Architects & Engineers (PAE) as regards infrastructure, and more in general the entire military-industrial complex engaged in the Vietnam War and the Gulf Wars.

Donald P. Green and Shapiro (1994), among others, have lessened the importance of the contribution of the School of Public Choice, which has remained a minority current in economic studies in recent decades, while Amartya Sen has written several articles in which he criticizes in particular the research of Buchanan and Tullock into collective political choices.

## Notes

1. If the weaker countries have been forced by their state of weakness to privatize violence, the United States has instead considered the use of contractors as a way to rationalize combat economically as well as a way to reduce their military presence in situations which public opinion would be against armed intervention to a large extent (cfr. Lizza 2001, p. 9).
2. Theodore Roosevelt, Richard Nixon and Henry Kissinger are among the "realists" in American politics because, differently from the Wilson's idealism, "they have to a certain extent seen the United States simply as a power among others", without attributing them particular messianic functions (Hassner and Vaïsse 2003, p. 57).
3. Waltz agrees with Schelling on this point.

4. This analysis, which achieved a success in the media above all after September 11 2001, has raised much criticism among researchers in various disciplines. A particularly pungent one came from Edward Said, an English specialist and critic of the concept of "orientalism", who published an article in *The Nation* on 22 October 2001 entitled *The Clash of Ignorance* which stated that labels such as "Islam" and "the West" served only to confuse ideas about the disorder in the contemporary world.
5. Giuseppe Bettoni, Italian geographer and follower of Yves Lacoste, shares the idea that geopolitics is not a science and states that the situation which may be analysed "with the method of geopolitics" has to have certain specific characteristics: the stake must be territory, two or more parties have to contend the stake, and a population must be involved in this contention (Bettoni 2009, p. 40).
6. Translation of *économie-monde*, taken in turn from *Weltwirtschaft*. The term was first used by Fernand Braudel and Immanuel M. Wallerstein.
7. Philippe Moreau Defarges, along with almost all French geographers who have taken an interest in geoeconomics, does not quote Luttwak's contribution.

## References


Bettoni, Giuseppe. 2009. *Dalla Geografia alla Geopolitica.* Milano: Franco Angeli.

Black, Duncan. 1948. "On the Rationale of Group Decision-Making." *Journal of Political Economy* 56: 23–34.

———. 1958. *The Theory of Committee and Elections.* Cambridge: Cambridge University Press.

Buchanan, James McGill. 1986. *Liberty, Market and State. Political Economy in the 1980s.* Brighton: Harvester Press.

Buchanan, James McGill, and Gordon Tullock. 1962. *The Calculus of Consent: Logical Foundations of Constitutional Democracy.* Ann Arbor: The University of Michigan Press.

Chiesa, Giulietto. 2000. "Il Tramonto della Geopolitica nell'Era Globalizzata?" In *I Dilemmi della Geopolitica e le Nuove Vie della Pace*, edited by Valter Collaruzzo. Milano: Guerini e Associati.

Falk, Richard. 1995. *On Human Governance.* Cambridge: Polity Press.

Giddens, Anthony. 1985. *The Nation-State and Violence: Volume 2 of A Contemporary Critique of Historical Materialism.* Cambridge: Polity Press.
Green, Donald Philip, and Ian Shapiro. 1994. *Pathologies of Rational Choice theory: A Critique of Applications in Political Science.* New Haven-London: Yale University Press.
Hassner, Pierre, and Justin Vaïsse. 2003. *Washington e il Mondo. I dilemmi di una superpotenza.* Bologna: Il Mulino.
Hoffmann, Stanley. 2002. "Le XXIe Siècle a Commencé." In *Vingtième Siècle.* Octobre-Décembre.
Jean, Carlo. 1995. *Geopolitica.* Bari: Laterza.
Kaldor, Mary. 1980. "Warfare and Capitalism." In *Exterminism and Cold War*, edited by Edward P. Thompson, et al. London: New Left Books and Verso.
———. 1981. *The Baroque Arsenal.* New York: Hill & Wang.
———. 1999. *New and Old Wars: Organized Violence in a Global Era.* Cambridge: Polity Press.
Kaplan, Robert David. 1994. "The Coming Anarchy." In *The Atlantic Monthly.* February.
———. 1997. *The End of the Earth, A Journey at the Dawn of the Twenty First Century.* London: Papermac.
Krugman, Paul Robin. 1991. *Geography and Trade.* Cambridge: Cambridge University Press.
Lacoste, Yves. 1976. *La Géographie, ça sert, d'abord à faire la guerre.* Paris: François Maspero.
———. 1993. *Dictionnaire de Géopolitique.* Paris: Flammarion.
Lizza, Gianfranco. 2001. *Geopolitica. Itinerari del potere.* Torino: Utet.
Lorot, Pascal. 1997. *Storia della geopolitica.* Trieste: Asterios.
Luttwak, Edward. 1999. *Turbo-Capitalism. Winners & Losers in the Global Economy.* London: Orion Business Books.
Mann, Michael. 1984. "Capitalism and Militarism." In *War, State and Society*, edited by Martin Shaw, 28–29. London: Macmillan.
———. 1988. *States, War and Capitalism.* Oxford: Basil Blackwell.
Moreau Defarges, Philippe. 1996. *Introduzione alla Geopolitica.* Bologna: Il Mulino.
Said, Edward W. 2001. "The Clash of Ignorance." In *The Nation*, October 22, n. 4.
Shaw, Martin. 1984. *War, State and Society.* London: Macmillan.
———. 1991. *Post-Military Society: Militarism, Demilitarization, and War at the End of the Twentieth Century.* Cambridge: Polity Press.

Solow, Robert Merton. 1993. "Blame the Foreigner." In *New York Review of Books.* December 16.

Waltz, Kenneth Neal. 1959. *Man, the State and War. A Theoretical Analysis.* New York and London: Columbia University Press.

———. 1973. "The Meaning of Anarchy." In *International Politics: Anarchy, Force, Imperialism,* edited by Robert Art and Robert Jervis. Boston: Little, Brown and C.

———. 1979. *Theory of International Politics.* Reading: Addison-Wesley.

———. 2008. *Realism and International Politics.* New York: Routledge.

Williams, Michael. 2007. *Realism Reconsidered: The Legacy of Hans Morgenthau in International Relations.* Oxford: Oxford University Press.

# 9

# Conclusions

The thought of economists concerning war has concentrated over time on different aspects and has greatly changed, while the subject matter to comment on has changed radically, seeing that the ways to wage war, the parties on the battlefield, the nature of clashes, the objectives announced from time to time to wish to continue with the use of violence have changed too.

If economic research faced historical reality, it would find it difficult to avoid the problem of war, because every historical period in the lifetime of each economist has seen warlike conflict of a greater or lesser intensity with its major impact on the economy, be it public or private.[1] Those who consider economics as a social science, therefore, face the reality of conflict, study its specific economic aspects and judge its consequences, often from an ethical point of view, too. But, if we leave aside the mercantilists and the Marxists, very few economists have recognized the sole, or sometimes the prevailing reason for war in factors connected with the economy, or in the mode of production. One of these economists is Gordon Tullock, whose research saw the common motivation and logic for unleashing violence, either individual or collective, and the search for a profit or the will to avoid a loss. He stressed the fact that his opinion was



R. Allio, *War in Economic Theories over Time*,
Palgrave Studies in the History of Economic Thought,
https://doi.org/10.1007/978-3-030-39617-6_9

not widely accepted because, while the defence (of the nation or private property) was usually considered as a rational act, aggression was usually considered to be the result of irrational factors. The desire to avoid losses was, therefore, held to be the typical cause of recent wars (Tullock 1974, p. 87).

Some economists have taken into account, often implicitly, an interpretation of the reasons for war put forward by economic historians and sociologists, which considers the economic causes of conflicts and, very briefly, limits them to two cases: shortage and overabundance. Raw materials shortages, for example, or a glut of capital compared to the opportunities for investment. This reading of events, generally speaking, is considered by its supporters as being valid over the long term.

The point of departure is to be found in the widespread agreement over the fact that in the past, when preindustrial societies were unable to expand production to obtain the requirements for maintaining the population, shortages, principally of reserves of food and raw material, but arable land too, provided the basic reason for wars. This motivation seemed to have been overcome thanks to the spread, at least in the western world, of the process of industrialization that allowed for the production of greater wealth than that which could be conquered at a much greater cost and with far greater risk through the booty that could be won with a victory on the battlefield.

They hoped for, and highly praised, peace achieved thanks to the political supremacy of the industrious bourgeoisie over the lazy nobility did not stand up to the depression at the end of the 1800s, and Europe therefore sought to establish new imperial frontiers, through colonialism, in order to obtain the raw materials which they continued to lack, or, more frequently, to export capital, goods or people that economic stagnation had made excessive for the national territory. Shortage and glut appeared side by side.

The 1900s saw the appearance of a new type of shortage, not of resources, but rather, of "vital space", adopted by German nationalism and chauvinism to unleash the most terrible of wars. Recent decades have seen the apparently paradoxical situation where some wars, which officially have been declared for various reasons, not infrequently "humanitarian" ones, were in fact wars to win or control scarce, or possibly dwindling, indispensable or strategic resources. Wars are also fought to seize various rare and indispensable minerals for new types of industrial production, which are either lacking or insufficient in highly industrialized countries. So, are we back to wars over shortages? It seems so, but they have to be fought using very expensive armaments in the field produced by major accumulation and an overabundance of capital.[2]

Secretary of State Alexander Haig made a declaration to the "Herald Tribune" in 1982, during the Reagan presidency, that we had entered "the era of resource wars", referring to the foreign policy of the Soviet Union being followed in Africa and the entry of Angola and Ethiopia into the Soviet orbit. Haig held that, if facts continued to follow that course, the USSR would come to control 90% of strategic mineral resources for which at that time there were no substitutes (cfr. Lizza 2001, p. 399).

Jeremy Black dedicated a book in 2004 to wars in the contemporary world and took up the question of war techniques. He concluded that the present-day interest in water, oil and basic resources in general recalled the struggles in a distant past.

> Pressure for water, oil and other resources lends specific intensity to what may also be a more generalized competition for living standards and jobs. While these factors may seem the product of economic change, population growth and environmental degradation in the modern world, they in fact, look back across the ages to competition over watering holes, grazing lands and the most fertile soil. The modes of conflict change greatly, as do their political, social, economic and cultural contexts but the root causes appear inherent to human society and they are readily apparent in conflict in parts of the world at present, although other factors also play a role. (Black [2004] 2007, p. 181)

The problem of the availability of water resources as the cause of conflicts breaking out has become more serious following climatic change over recent years, which has led to the desertification of areas short of water even beforehand. The result has been mass migration, tension between religious and social groups and, of course, war.[3]

Two French summaries, quoted above, by Gaston Bouthoul and Armelle Le Bras-Chopard provide a short consideration of war as a consequence of shortages or overabundance. Bouthoul, in particular, reminds us that war in contemporary society is a luxury activity as unleashing it requires having previously accumulated a consistent surplus to be invested in armaments which have, in the meanwhile, become ever more sophisticated and expensive, even if, as Marshal Saxe had stressed back in the early 1700s, war always needed a large amount of money. Consequently, it would always be overabundance and not shortage, usually officially considered the cause, that makes wars break out in the modern world.

As has been previously stated, overabundance was the cause of the colonial wars in the 1800s, at least of those used to seek outlets for production in excess of national demand or for capital investments for excess capital: wars of the rich against the poor in which, in fact, as Bouthoul comments, the opposite to what Marx had forecast came to pass.

Marxists, as is well known, see wars as the inevitable result of the internal contradictions of the capitalist system, but according to Rosa Luxemburg the capitalist economy seeks to overcome these contradictions by seeking territory outside the capitalist mode of production to find a way to get rid of products left unsold on the home market. This led to the race for colonies by the capitalist industrial states. Lenin instead considered that it was excess capital that sought new frontiers.

Hannah Arendt too thought that imperialism was a product of an overabundance, or rather two: of savings and also people, who, like money, did not find suitable employment in the West during periods of economic slowdown.

The theme of shortage and overabundance drew the interest of sociologists above all, and it was Aron who stressed that the reality of shortage was of little importance in unleashing wars, while the perception of it was fundamental. The belief in the existence of shortage fed by warmongers, or the trust, again something invented, in the possibility to obtain significant advantages from war, bolstered warlike doctrines and the wish for conquest. This holds true for the shortage of land complained of in the twentieth century through the concept of the need for "vital space" for aggressive, expanding populations. What really counted then in terms of economic growth was not so much material space, but rather the technology and innovation, the perception of a shortage on the part of the masses was consciously channelled into territorial questions in order to make wars break out. In any case, Aron thought, the theory of shortage could no longer be evoked in contemporary society, because preparations for war today require a considerable surplus or accumulation.

The list of those who agree indirectly with the fact that overabundance causes armed conflicts between states also includes those economists who consider that the function of waste carried out by war is useful, in fact, indispensable. War is truly able to rapidly destroy the stocks of useless goods that have accumulated in a country, that is to say in economic terms, that war effectively writes off excess products. Furthermore, this sterilization takes place under the direct control of the state. Therefore, a factor such as destruction, even if in itself is negative from an economic point of view, can be useful in straightening out the accounts. This thesis is expanded in a cynical, but persuasive manner in the anonymous pamphlet quoted above, presumably written by Galbraith, concerning the undesirable nature of perpetual peace for the economy in the present states of affairs.

Nevertheless, such a complex phenomenon as war can hardly have a single cause. Gaston Bouthoul, who proposed the creation of a science of war he called polemology, recalled that wars are *polytéliques,* that is, they always have different causes and a multiplicity of ends. He stated that one of the few wars that were openly presented as a war for an economic reason was the Opium War between China and Britain. For Bouthoul, who sought to understand the possibly recurrent necessity for wars and what could be the various manifestations that they led to, warlike conflicts are in fact a sign of instability in a social, rather than economic, system, within which a large number of variables made political and military control illusory. War from this point of view is therefore not a system, but the manifestation of a systemic dysfunction within which conflict has exploded (cfr. Schmidt 1991, p. 133).

While Clausewitz saw war as a public activity founded on the state, the army and the people, nowadays we can see an often changing and complex world of "Warlords" that also includes "criminal gangs, non-government organizations and international networks (like al Qaeda) engaged in the use of violence" (Collaruzzo 2003, p. 57).

There are, however, very few economists who seek to evaluate the role of these agents and of arms-producing industries in unleashing wars and sharing out the booty. In the same way, it is not economists, but missionaries, politologists and students of international relations who enquire into the operation of corruption undertaken by multinationals, which thereby disarticulates local political power in underdeveloped countries where useful mineral resources are located.

If we look at interpretations of the relationships between militarism and capitalism, leaving aside the Marxists' one, there are three main cases. One theory holds that, even if militarism was already widely developed in pre-capitalist societies, capitalism reinforced it further and more powerfully, thanks above all to technological innovation in the armaments sector. An optimistic theory, falling within the old free trade framework, still sets out from the assumption that industrial capitalism is eminently pacific and will, sooner or later, after a two-century wait, favour the decline of militarism. A third theory, that bases itself on geopolitics and neorealism, holds that militarism is basically due to political reasons which are historically permanent and linked to aspects typical of relationships between states or territorial questions, which have seen few changes, except for the technologies used in war production.

This is the case for economists who display an interest in the problems posed by economic reality, and thus are also interested in war, above all when significant warlike phenomena are underway.

Classical economists from Smith on showed, if anything, only an incidental interest in war. They trusted in the fact that the development of productive techniques, by leading to the satisfaction of the population's needs, would make war useless. The problem of the defence of the state remained, but it was a political, not an economic one.

Neoclassical economists, marginalists, the supporters of the all-healing values of the free market and theoretical economists in general, who consider economics to be an exact science, are not interested in war. Instead of analysing the economic problems posed in the real world, the economy as it is, they prefer to study the economy as it should be, that is, by concentrating their attention on the solution to problems that would require, in the first case, situations of peace in order to be resolved. They construct econometric models and establish universal eternal laws to find optimal solutions to individual problems isolated from the context of economic reality, assumed to be fixed conditions, which, in reality, are not fixed at all. They pose the question as to which economic conditions and which types of behaviour could lead to the realization of the equilibrium between supply and demand, the optimal distribution of available resources and maximization of the utility function. These economists, as we have seen, do not include the state of war among the conditions given, as it appears to be unsettling and does not obey eternal universal laws. The neo-institutionalist economists D. C. North, J. J. Wallis and B. R. Weingast published a book in 2009 on violence and social orders. They stressed how the systematic avoidance of an analysis of conflicts had led economists to underestimate the "unproductive weight" of violence in their calculations. They therefore hoped for the elaboration of new theories that would be able to integrate the existence of violence in economic models:

> the existing body of knowledge in social science can be transformed by a new conceptual framework that changes the way we think about traditional problems in economics, political science, sociology, anthropology, and history that result from an explicit consideration of the role that violence plays in shaping social orders, institutions, and organizations and their development over time. Our recommendations for new research entail an in-depth understanding of violence, institutions, organizations, [...] that we do not currently possess. (North et al. [2009] 2013, p. 271)

Nevertheless, back at the end of the 1980s and the beginning of the 1990s, Jack Hirshleifer had already developed a model to provide the results of production choices and the allocations of resources in both productive and non-productive activity when conflict was underway.

In the same period, certain mathematical economists, in fact war strategists, set up an "economic theory of conflict" using Game theory in their analysis and used it to define the rationality of war based on economic criteria. Aumann was the leading representative. He won the

Nobel Prize for economics, not without criticism, and in his acceptance speech he, not by chance, did not speak of economic questions, but of war and peace and considered the choice of war as rational as it corresponded with the criteria established by Walras.

Economists, who are convinced that mankind makes choices in a rational way concerning their greater interest, can adopt different positions when facing war, which, in itself, appears as the most irrational human activity, as well as the most uncertain in its outcome. All the same, the choice to unleash it or not does not fall to citizens, who have to accept the consequences, or to men who fight it; instead, it falls to politicians and great electors, who have an economic interest to start conflicts. Therefore, following their decision, the choice to unleash war could be economically rational, even when it is morally deplorable.

The analysis Aumann made of war is theoretical; the science, which is the object of his research, as it is called, meets the criteria of exact science and, as such, is not interested in the practical consequences of what has been discovered and the results obtained are not subject to judgement of a moral or ethical kind. Therefore, the hypothesis of atomic war with millions of dead and the destruction of the environment did not worry Aumann or any other economists working on war strategy, who calculate the most opportune way to conduct conflicts using the rules for playing poker. The same holds true for the economists of Public Choice.

The students of Game theory, moreover, do not want war unless it is rational and bet on deterrence to ward it off, even though Luttwak, certainly no pacifist, correctly remarks that, with the actual costs of armaments and military research, deterrence can only be afforded by strong states.

The theoreticians of games rationally calculate the best military strategy to adopt in case of conflict. Besides, in the case that the choice of war is held to be rational and, considering economics as an exact science, the theme of violence can be taken up without difficulty, because exact sciences do not take into account the consequences of what they discover.

> Our strategists are also exactly like the other people of their social class, education, and background. They enjoy the same sort of personal relations as the rest of us, appreciate the same gifts that the life bestows, suffer from the same grief and misfortunes. The monstrosity of their work carries little or no emotional meaning for them, not because they are mentally ill, but because they share with the rest of us or perhaps are more richly endowed than the rest of us with the most creative of human faculties, which becomes also the most dangerous one when coupled with a lack of extensional imagination – the faculty of abstraction. (Rapoport 1964, p. 192)

Mathematical economists are not mentally ill, of course, even if the case of von Neumann, who in the immediate post-war period suggested that the United States should make a preventive atomic attack of the Soviet Union, gives rise to doubt. They are mentally sane, but, when they consider the wars of their time, they are often warmongering. Apart from Neumann, an enthusiast of the atom bomb, Aumann is noted for his bellicose stand on the Israel–Palestine problem.

Rapoport instead, even if he used instruments of theoretical analysis, turned against warlike conclusions in the name of a "basic" ethic which, according to him, transcends and supersedes economic interest and the economic rationality of war. However, Rapoport was not just a mathematician, but also a psychologist and stated that he was repulsed by the strategists financed by the Pentagon who, during the Cold War, calculated the acceptable cost in human lives when using the atom bomb. Thus, Kuznets, who did much work of the definition of costs in the Second World War, found difficulty in translating the horrors of war into dollars and cents.

Obviously, economists who conduct research into concrete reality are sensitive to the human implications of war. They consider economics as a social science in which human actions can be subject to an ethical judgement. To offer just one example, apart from the evaluations given of Rapoport and Kuznets, the last Tanner Lectures on Human Values held by Joan Robinson "was a criticism of the nuclear arms race in which she argued in terms of economics, but also appealed to the audience's morality" (Harcourt and Kerr 2009, p. 12). In the lecture, Robinson made clear her vehement condemnation of warmongers and quoted the anti-nuclear activist Helen Caldicott in order to attack the mindlessness of MAD:

> Countries driven by fear and a mutual distrust bordering on the pathological, are locked into a suicidal strategy calling, in the world of the Pentagon, for 'mutually assured destruction' (MAD) as the best deterrent to war. But 'arms for peace' and 'security through mass genocide' are strategies that defy logic and common sense. They epitomize our nuclear madness. (Robinson 1982, p. 251; Caldicott 1979, p. 17)

Then, of course, there have been, and even more so, there are today, economists whose research work explicitly seeks solutions for disarmament, peace and cooperation. Universities in various countries have in recent years begun to offer courses in peace economics, even if under other names, which seek to establish measures in economic policy that may be able to remove the causes of armed conflicts. Peace economics overturns the framework of traditional economists of a liberal background who generally see economic development as the necessary condition for realizing long-term international peace. Peace economics, on the other hand, holds that a situation of lasting peace is required in order to be able to achieve harmonious economic growth.

Economists who fight for peace are not just a few individuals and are often highly qualified and in most cases work together in associations, either exclusively composed of economists, or interdisciplinary in membership. The two longest-lived societies are the "Peace Science Society International" and the "Economists for Peace and Security". The former was established in Oslo in 1963 by Walter Isard, an American economist, and has accepted members from various social disciplines from the outset. It started publishing the "Journal of Peace Research" in 1964, which still appears. Isard in fact conceived "Peace Science" as a full-blown multidisciplinary autonomous science of an international character interested in the study of the reasons for violence and conflicts, their effects and their possible solutions. Over the years, the presence of economists has continued to decline proportionally, falling below that of scientists from other disciplines whose research, according to Isard, was essential for economists when adapting their theoretical analyses to the complex world of reality (Isard and Anderton 1992, p. XV).

The "Economists for Peace and Security" is an association which has changed its name several times during its history. It was founded in 1989 as "Economists Against the Arms Race", becoming "Economists Allied for Arms Reduction" in 1993, adopting its present name in 2005. It is an international non-government organization of economists located in New York. The association was founded by two American Nobel Prize-winning economists: Kenneth Arrow and Klein Lawrence. Other Nobel Prize winners are among its members, including Amartya Sen and Oscar Arias (Nobel Peace Prize). James Kenneth Galbraith, son of John Galbraith, was named Chair of the Board of Direction in 1996 and Intriligator was Vice President. It is involved in theoretical research and policy development aimed at the control of arms proliferation worldwide. The English section publishes a review called "The Economics of Peace and Security Journal".

Many other national and international organizations have been set up over recent decades to bring together scientists who often come from different disciplines, always including economics, who work together to establish and develop peace economics. Many of these associations publish their own reviews, mainly online.

Economists from various countries take part in the associations "Scientists for Peace", "Global Union of Scientists for Peace", "Institute for Economics and Peace", "Network of European Peace Scientists", "Italian Union of Scientists for Disarmament" and others as well.

The research and effort made by these militant economists go to confirm what Karl Deutsch (1912–1992) wrote: "war, to be abolished, must be understood. To be understood, it must be studied".

And studies on war must be published because

> the scientist by himself can neither start nor stop a war [...], but a scientific analysis of the causes of war, if convincing to the people at large, could be one effective as well as a democratic force for peace. We have to make it clear to the common people of the world that any aggression anywhere is, in the last analysis, war against them.

The Indian indologist and pacifist Damodar Dharmananda Kosambi wrote this in 1951. Unfortunately, after nearly 70 years the voice of scientists continues to be drowned out by the scream of unpunished lies of war.

## Notes

1. "Guerra è sempre" ["There's always war"] states Mordo Nahum, the Greek friend of Primo Levi in his novel "La Tregua" ["The Truce"] when he is informed that the Second World War is over.
2. Lizza (2011, p. 22) has observed that recently the race, first colonial, then imperialist, to grab energy sources has been faltering, while the strategic interest of producers and consumers of oil and gas resources has focused on the oil and gas pipelines that criss-cross the world.
3. The IHE Delft Institute for Water Education of UNESCO foresaw that the shortage of water would be an important factor in unleashing conflicts in the future. It started a programme entitled "Water, Peace and Security" (WPS), in order to understand where battles could take place over water shortages and how to seek to prevent them. The programme analyses the growth and density of the population, precipitation, the type of infrastructure and the characteristics of the government.

## References

Black, Jeremy. [2004] 2007. *War Since 1945*. London: Reaktion Books.

Caldicott, Helen (with the assistance of Nancy Herrington and Nahum Stiskin). 1979. *Nuclear Madness: What You Can Do!* New York: Autumn Press.

Collaruzzo, Valter. 2003. "Nuovi nomi per nuove guerre." In *Guerre globali. Capire i conflitti del XXI secolo*, edited by Angelo d'Orsi. Roma: Carocci.

Harcourt, Geoffrey, and Prue Kerr. 2009. *Joan Robinson*. London: Palgrave Macmillan.

Isard, Walter, and Charles H. Anderton, ed. 1992. *Economics of Arms Reduction and the Peace Process: Contributions from Peace Economics and Peace Science.* Amsterdam: North-Holland.

Kosambi, Damodar Dharmananda. 1951. "Imperialism and Peace". *Monthly Review* 3, June 2.

Levi, Primo. 1963. *La tregua.* Torino: Einaudi.

Lizza, Gianfranco. 2001. *Geopolitica. Itinerari del potere.* Torino: Utet.

———. 2011. *Geopolitica delle prossime Sfide.* Torino: Utet.

North, D. C, J. J. Wallis, and B. R. Weingast. 2009. *Violence and Social Orders, A Conceptual Framework for Interpreting Recorded Human History.* Cambridge: Cambridge University Press.

Rapoport, Anatol. 1964. *Strategy and Conscience.* New York: Harper and Row.

Robinson, Joan. 1982. "The Arms Race." In *The Tanner Lectures on Human Values.* Vol. III. Salt Lake City: University of Utah Press.

Schmidt, Christian. 1991. *Penser la guerre, penser l'économie.* Paris: Édition Odile Jacob.

Tullock, Gordon. 1974. *The Social Dilemma: The Economics of War and Revolution.* Blacksburg, VA: University Publications.

# Index

A
Adams, Gordon 127
Aeronautics 188, 199
Afghanistan 8, 105, 150
Africa 30, 94, 198, 227, 247
Agreement 145
cooperative, 154
al Qaeda 250
America 105, 107, 109, 112
American Department of Defense 8
Amin, Samir 91
Anarchy 87, 159, 218, 228
consequent, 222
future, 227
of the system, 221, 228
*Ancien Régime* 77
Anderson, Perry 73, 74, 98
Anderton, Charles H. 255
Angell, Norman 165, 220
Angola 207, 247
Anonymous (William Petyt?) 18
Ansaldo, company 182
Arendt, Hannah 93, 94, 98, 99, 248
Argenson, René-Louis de Voyer de Paulmy Marquis de 19
Arias, Oscar 255
Arm 26, 40, 113, 197. *See also* Armament(s)
absolute, 223
atomic, 152
common, 223
importance of, 235
nuclear, 223
race, 155, 158, 161
sophisticated, 235
trade, 159
Armament(s) 3, 4, 16, 26, 36, 38, 47, 49, 60, 72, 79, 80, 106, 107, 110, 112, 113, 116, 139, 150, 156, 175, 178,


R. Allio, *War in Economic Theories over Time*,
Palgrave Studies in the History of Economic Thought,
https://doi.org/10.1007/978-3-030-39617-6

179, 207, 209, 250. *See also* Disarmament
control of, 194, 201
cost of, 121, 179, 180, 193, 209
demand for, 84, 112, 190, 208
development of, 234
escalation of, 210
expenditure on, 79, 80, 84, 109, 115, 134, 140, 226
export of, 190
function of, 227
huge, 152
impact of, 135, 165
increase in, 176, 192, 203, 208, 209
innovation in, 104
investment in, 158
level of, 115
modern, 228
nuclear, 135, 153, 194, 199, 205, 210
production of, 79, 84, 112, 225
purchase of, 238
resources for, 135
sale of, 178
supply of, 134
technological spin-off of, 176, 177, 179–181
wasted on, 106
Aron, Raymond 3, 5, 46, 47, 49, 50, 92, 144, 145, 219, 248, 249
Arrow, Kenneth 255
Asia 198
Southeast, 235
Atavism 70
form of, 71
Atlee, Clement 121
Atomic Energy Commission 195, 198
Aumann, Robert John 148–150, 152, 164, 251, 252

B

Bacon, Francis 14
Baltimore 198
Bank of America Corporation 108
Baran, Paul Alexander 184, 190, 191
Barnet, Richard Jackson 194
Bastiat, Frédéric 40, 41, 44, 239
Baudrillart, Henri 41, 42
Beccegato, Paolo 11
Behaviour 25, 44, 70, 92, 114, 125, 137, 138, 144, 155, 157, 190, 251
economic, 236, 238
electoral, 238
human, 221
political, 221, 236, 238, 239
Belgium 62, 202
Belgrade 191
Bell Laboratories 148
Benin 207
Benoit, Émile 189
Bentham, Jeremy 24
Berlin 124, 217, 231
Berlin Wall 217, 231
Bernholz, Peter 145, 147
Bettoni, Giuseppe 231, 236, 242
Bientinesi, Fabrizio 210
Bilmes, Linda J. 105–107
Bipolarism 5, 6
Bipolarity 223
end of, 223
Black, Duncan 238, 239
Black, Jeremy 247
Blair, Anthony Charles Lynton 149
Bodin, Jean 15, 17, 18

Boffa, Giuseppe 206, 211
Boisguilbert, Pierre Le Pesant Sieur de 16
Boldizzoni, Francesco 68
Bolivia 207
Bond 121–123
  wartime, 116
Border 9, 140
  abolish, 217
  established, 217
  within, 217, 225
Borel, Emile 137
Botero, Giovanni 15
Boulding, Kenneth E. 147, 156, 157, 189
Bourgeoisie 77, 93, 94, 98, 246
  behaviour of, 73
  industrial, 70, 72
  rise of, 70
Bouthoul, Gaston 156, 161, 248, 249
Braithwaite, Richard B. 142
Braudel, Fernand 232, 242
Brezhnev, Leonid Ilyich 202
Britain 249
Buchanan, James McGill 238–241
Bukharin, Nikolai Ivanovich 82, 86–88
Burckhardt, Jacob 4
Burns, Arthur Frank 108, 109, 198
Bush, George Walker 6, 8, 9
Buzuev, Aleksandr Vladimirovich 207–211

C

Cabiati, Attilio 102, 122
Caldicott, Helen 254
Cambridge 102
Cannan, Edwin 159, 160
Capital. *See also* Capitalism
  accumulation of 64, 74, 84, 85, 88, 91, 94, 186, 228
  circuit, 122
  competition between, 225
  concentration of, 197
  destruction of, 109
  diversion of, 107, 113
  domestic, 95
  domination of, 83, 86
  excess, 68, 72, 248
  export of, 82, 88–90, 97
  finance, 67, 82, 83, 85–89
  French, 96, 97
  increase in, 122
  industrial, 81, 86, 87, 89
  internationalization of, 217, 225
  investment of, 186, 187, 196, 207
  military, 103
  national, 225, 228
  need of, 74
  private, 95, 96
  Russian, 96
  shortage of, 96
  superfluous, 94
Capitalism
  crisis of 201
  development of, 64, 70
  European, 84
  finance, 67
  form of, 82
  imperialist, 49
  inadequate for, 67
  industrial, 86, 88, 163, 225–227, 250
  interest of, 66
  late, 185

military, 186, 191, 194, 197, 200, 202
monopoly, 172, 185, 201
nature of, 73
organized, 87
peaceful, 45, 49
presence of, 225
private monopoly, 88
realignment of, 66
stage of, 70
state, 161, 163, 172, 191, 194, 195, 201
state monopoly, 88
twentieth century, 74
useful to, 68
Capitalist 81–84, 86, 87, 89–92, 95, 98
Carnegie Foundation for International Peace 101, 125
Cartel 81, 86
Caruso, Raul 163, 164
Center of Systemic Peace 4
Charter of the United Nations 101
Chauvinism 73, 246
Cheney, Richard Bruce 8
Chevalier, Michel 41, 42
Chiesa, Giulietto 237
Chile 207
China 178, 249
Chomsky, Noam 126
CIA 193, 194
Clarke, Richard William Barnes 103
Clark, Grover 102
Clark, John Maurice 102
Class 32, 43, 45, 47, 226, 253
antagonism of, 77
reactionary, 86
ruling, 69–71, 73
social, 72, 73
struggle of, 77, 84, 90
upper, 81, 94
war-making, 73
working, 84, 85, 89, 90
Clausewitz, Carl von 150, 155, 249
Clemenceau, Georges 117
Clinton, William Jefferson 106
Cobbett, William 30
Cobden, Richard 26, 42
Coercion 151
Coexistence 203
peaceful, 203, 205
Cohen, Richard 127
Colbert, Jean-Baptiste 16
Cole, George Douglas Howard 103
Collaruzzo, Valter 250
Colombia University 162
Colonialism 25, 43, 65, 82, 246
British, 78
causes, 88, 95
financial, 81
French, 96
need for, 96
neocolonialism, 78
Colonies 79, 81–84, 86, 94, 96, 98
British, 81
Colonization 26, 27, 29, 32, 33, 51
Combat 149, 151, 154
economic formalization of, 136
Communism 6, 175, 200, 202, 205, 206
Company
armament 180, 190, 196, 199
European, 182, 190
industrial, 182
military, 182, 190, 195, 196, 198
private, 218
public, 182, 196
state, 190, 195

Comte, Auguste 46, 78
Conflict
analysis 140, 145
approach to, 145
armed, 74, 189, 204, 207, 223, 249, 254
behaviour, 152
cause of, 219, 221, 226
consequence of, 157
cost of, 105, 151, 152
domestic, 140
economic role of, 172
emergence of, 218
foreign, 140
global, 123
ideological, 172
international, 58, 60, 65, 139, 152, 157, 177, 178, 224, 235–237
Iraq, 106
irreconcilable, 142
Israel–Palestine, 236
limited, 154
means of, 231, 234
new, 218, 233
object of, 151
of interest, 189
permanent, 201
persistence of, 224
problem of, 230
reality of, 245
reason for, 142, 228
short-term, 105
social, 171
solution to, 156, 157
strategy of, 153
territorial, 235
theme of, 59, 64
warlike, 1, 4, 6, 62, 68, 73, 133, 148, 156, 157, 159, 160, 245, 249
Congo 7, 207
Consumption
capacity of 123
civil, 119, 121
deferred, 119
family, 125
increase in, 119
limitation on, 121
private, 109
Contract 137
Cooperation 222, 236, 254
effective, 224
international, 219, 220
Corporation
American 196, 201
military, 196, 208
private, 195
security of, 196
Cost
analysis 101
calculation of, 155
comparative, 102
competitive, 58
direct, 101, 105, 109
evaluate, 150
indirect, 101, 105, 109
invisible, 108
living, 119
low, 68
marginal, 134
material, 105
minimization of, 133, 151
monetary, 116
of war, 101, 113, 114, 116
opportunity, 104, 107, 109, 111, 113, 116, 179–181, 198

overall, 103
production, 57, 187
useless, 60
Country 224, 229, 235–237, 241
developed, 218
underdeveloped, 235
Cournot, Augustin 136, 154
Cramer, Christopher 147, 148
Crisis
Cuban Missile Crisis 154, 164
deep, 174
economic, 4, 171, 175, 187
social, 171
Suez crisis, 150

D

Davenant, Charles 15, 114
De Arcangelis, Mario 185
Debt 105, 133, 240
inter-ally, 118
long-term, 114
public, 60, 61, 106, 112, 114, 115, 122
war, 115
Defence
administration 195
character of, 240
department, 192, 195, 196, 198, 208
externality of, 157
federal, 199
industry, 228
lines of, 232
military, 193, 196, 197
ministry of, 235
national, 175
organization of, 146
problem of, 135
programme, 135
requirement of, 218
De Finetti, Bruno 136, 142, 143, 145
De Grasse, Robert 127
Delft Institute for Water Education 256
Demand 30, 37, 38, 64, 66, 68, 81, 111, 114, 118, 120, 251
increase in, 171, 174, 208
military, 186, 189
overall, 171
public, 174
state, 182, 190, 208
strong, 174
unproductive, 190
de Molinari, Gustave 41, 42
Denmark 43
Department of State 106
Desiderio, Alfonso 10, 174
Destruction 171, 172, 184, 192, 194
means of, 185, 186
Deterrence
effect of 192, 194
effectiveness of, 224
policy of, 154
power of, 194
terms of, 135
Deutsch, Karl 255
Development. *See also* Underdevelopment
capitalist 68
destabilizing, 222
economic, 2, 8, 60, 61, 66, 175, 177, 183, 198, 203, 209
industrial, 59, 60, 172, 177, 183–185, 225
laws of, 66
level of, 235

long term, 186, 207
process of, 58, 63
technological, 172, 177, 178, 184, 227, 234
de Viti De Marco, Antonio 239
Di Nolfo, Ennio 6
Disarmament 135, 163, 191, 254
choice of, 189
controlled, 205
total, 204, 205
unilateral, 208
Doctrine 191
Truman, 201, 202
Zhdanov, 201
Domhoff, William George 193
Downs, Anthony 239
Dunoyer, Charles 41
Dupont, company 198
Dupont de Nemours, Pierre-Samuel 27, 35
Durbin, Evan Frank Mottram 103, 121
Durkheim, Emile 47, 49

E

Eastern Europe 172, 202
East India Company 24
Econometric Research Program 164
Economic balance 19
Economic(s)
activity 16, 19, 23, 32, 51, 154, 163
analysis, 40, 42, 44
applied, 135
competition, 135–137
competitiveness, 26, 27
conception of, 58
cooperation, 34
*cosmopolitical*, 59
damage, 140
domination, 17, 28
equilibrium, 44
evolution, 28, 41
expansion, 84, 93
fallout, 158
freedom, 27, 35
growth, 23, 32, 33, 162, 163
interest, 81, 92, 96
laissez-faire, 57, 58, 60, 65
law of, 36, 57, 94
mean, 20, 23, 28, 29, 40
modern, 151, 164
national, 59
necessity, 159
neoclassical, 147
opportunity, 159
peace, 249, 254, 255
policy, 13, 32, 40
power, 13, 14, 19, 32, 40, 47, 82, 91
progress, 25, 32, 36, 38
prosperity, 26, 31
rationality, 147
reasons, 81, 82, 85, 92
reflection on, 64
resource, 14, 42
school of, 62
science, 13, 25, 26, 36, 40
settlement, 137
stagnation, 31, 32
supremacy, 22
system, 27, 40
theory, 13, 14, 77, 136, 137, 148, 159
utility, 147
Economist 3, 4, 18, 22, 27, 28, 30, 33–35, 42, 46, 57, 59, 64, 69,

91, 103, 121, 134, 138, 145, 148, 156, 158–160, 162, 172, 175, 200, 207, 245, 250, 251, 254
association of, 158
classical, 133
free trade, 5
marginalist, 3, 4, 136, 250
mathematical, 133, 155, 251, 253
monetarist, 4
neoclassical, 3, 4, 250
neo-institutionalist, 251
theoretical, 135, 157
Economists Against the Arms Race 255
Economists Allied for Arms Reduction 255
Economists for Peace and Security 254, 255
Economy 2–5, 245, 250. *See also* Economic(s); Economist
agricultural, 60
American, 103, 107, 108, 173, 190, 192, 194, 196
arms, 173, 186, 187, 191, 203, 207
British, 36, 38, 180
capitalist, 83, 90, 93, 217, 226, 248
civil, 158, 161
civilian, 179, 184, 185, 207
cosmopolitanism of, 58
dependence of, 124
development of, 60, 68
domination of, 217
free trade, 225
global/globalized, 91, 163, 217, 222, 227, 233, 234, 237
impoverishment of, 105
industrial, 60
intervention in, 58, 69, 101, 235, 238
law of, 34, 36
market, 35, 45, 92
military/militarisation of, 138, 162, 174, 180, 183–188, 193, 194, 198
national, 59, 88, 93, 135, 221, 234
of scale, 151, 155
pacifism of, 58
peacetime, 104
planned, 69
political, 25, 39, 41, 42, 44, 59, 60, 161, 235
public/state, 36, 240
real, 233
war, 13, 34–36, 39, 41, 42, 69, 158, 160–163, 172–174, 186, 187, 189, 193
wartime, 102, 112
world, 110, 225, 233
Edelstein, Michael 104, 109
Edgeworth, Francis Ysidro 136, 137, 164
Egypt 73, 97
Eichholtz, Dietrich 172, 200, 201
Einaudi, Luigi 5, 101, 125
Einzig, Paul 103
Eisenhower, Dwight David 192, 193, 207
Electronics 185, 188, 198, 199, 208
Empire 115
Employment 188, 190, 198, 205, 209
Enfantin, Prosper 51
Engels, Friedrich 77, 78
England 28, 61

Enthoven, Alain 134
Entrepreneurship 196
Equilibrium 3, 118, 136, 141, 251
  general, 136
  irrationality of, 155
  social, 146
  strategic, 150
Ethiopia 247
Europe 6, 17, 18, 21, 30, 37, 43, 45, 48, 62, 65, 66, 87, 101, 111, 125, 177, 183, 185, 202, 207, 246
  Central Europe, 118
European unification 5
Exchange 136, 146
Expansion 6, 24, 46, 176, 240
  industrial, 173
  territorial, 172
Expansionism 80, 95–97
Expenditure 135, 162
  American, 107
  armaments, 178, 180, 185, 187, 189, 190, 192, 196, 203, 209
  defence, 103, 104, 106, 175, 198
  exorbitant, 109
  indices for, 103
  military, 107, 111, 112, 140, 150, 157, 175, 179–181, 184, 185, 187–189, 191, 192, 195, 196, 198, 199, 209
  military/militarisation of, 188
  necessary, 104
  of war, 175, 179, 184, 187–189, 191, 196, 199
  on arm, 115
  on peace, 115
  on security, 115
  parasitic, 196
  public, 180
  requirements to, 115
  unproductive, 116, 179, 190, 196
  war, 4, 104, 106, 114, 115, 125
Exxon-Mobil 107

F

Falk, Richard 227
Fascism 176, 201, 231
FBI 193
Ferrara, Francesco 239
Ferry, Jules 96, 97
FIAT (company) 182
Fiesole 183
Finance 217, 233, 234
  state, 217
Financing
  federal 108
  means for, 102, 109
  ways of, 114, 121
Fontanel, Jacques 175
Force 26, 31, 49, 81, 83, 106, 111, 115, 123, 136, 157, 207, 224, 255
  armed, 185, 195, 198, 211
  political, 204
  productive, 173
  reactionary, 202, 205
  use of, 191
France 19, 20, 23, 39, 41, 73, 77, 96, 97, 115, 161, 165, 182, 183, 202, 209, 218, 230, 231
Freedom 193
Free trader 2
Freiburg 63
Fridenson, Patrick 181, 182
Friedman, Milton 126
Fukuyama, Francis 237

G

Gain 21, 25, 27, 29, 30, 32, 35, 71, 96, 124, 151, 240
  easy, 173
  economic, 106, 172, 200
  illegal, 106
Galbraith, John Kenneth 188, 189, 193, 195, 196, 255
Galiani, Ferdinando 18
Gallagher, John 28
Game 17, 80, 142
  cooperative, 154
  non-cooperative, 138
  non-zero-sum, 139, 154
  qualitative, 154
  quantitative, 154
  repeated, 150, 154
  zero-sum, 137, 139, 153, 154
Game theory 1, 3, 4
Gaza Strip 164
General Dynamics 241
General Electric, company 198
General Motors, company 198
Genovesi, Antonio 28
Geoeconomics 218, 231–233
  emergence of, 232
  era of, 234
  instrument of, 234, 235
  label of, 232
Geopolitics 250
  domain of, 230
  economic, 218, 232, 236
  method of, 230
  range of, 232
Germany 23, 48, 62, 64, 67–69, 81, 82, 92, 96, 115, 118, 172, 175, 183, 201, 230
Gerschenkron, Alexander 74
Giddens, Anthony 224, 226, 227
Gilling, Philippe 51
Globalization 8
  cultural, 237
  economic, 222, 227, 237
  effects of, 224, 237
Global Union of Scientists for Peace 255
Goods 21, 27, 40, 44, 63, 81, 83, 107, 109, 110, 112, 118, 120, 146, 162, 188
  circulation of, 217
  consumption of, 217
  production of, 217
Government 16, 31, 32, 59, 105, 115, 124, 186, 190, 195, 198, 208
  American, 236
  behaviour of, 221, 236, 238
  cosmopolitan, 227, 228
  form of, 221, 227
Great Britain 16, 20, 23, 27, 32, 34, 37, 43, 45, 48, 51, 57–59, 62, 64, 66, 79–81, 94, 96, 98, 115, 118–120, 122, 140, 183, 188, 209, 228, 230
Greece 202
Green, Donald Philip 241
Green, Murray 73
Gross Domestic Product (GDP)/Gross National Product 8, 103, 104, 109, 112
Growth 218, 226, 236, 256
  capitalist, 64, 66
  condition for, 64
  disastrous, 193
  economic, 62, 175, 176, 181, 183, 190, 191, 194, 202, 206, 249, 254
  evaluation of, 181, 198

lack in, 108
long-term, 103
of population, 189, 209
post-war, 104, 174, 185
rate of, 108
Guatemala 207
Guibert, Jacques-Antoine-Hippolyte Comte de 113, 127

H

Habermas, Jürgen 210
Haig, Alexander 247
Hallgarten, Georg Wolfang Felix 199
Halliburton 241
Hamilton, Walton 102
Harcourt, Geoffrey 253
Hart, Sergiu 148
Harvard 97
Harvey, David 126
Hassner, Pierre 9, 224, 229, 241
Haushofer, Karl 229, 230
Hawtrey, Ralph George 91
Hayek, Friedrich August von 120
Heidelberg 63
Higgs, Robert 109
Hilderbrand, Bruno 62
Hilferding, Rudolf 82, 83, 85, 87, 89, 90
Hiroshima 152, 184
Hirshleifer, Jack 145–147, 158, 251
Hirst, Francis Wrigley 111, 115
Hitch, Charles 134
Hitler, Adolf 103, 124, 172
Hobbes, Thomas 219
Hobsbawm, Eric John 98
Hobson, John Atkinson 2, 71, 78–81, 85, 92, 97
Hoffmann, Stanley 237, 238
Holland 16
Hont, István 16, 24
Huet, Pierre-Daniel 17
Human Rights 227
Declaration of, 101
Hume, David 23, 24
Hungary 62
Huntington, Samuel Phillips 9, 227–229

I

Idealism 219
opposition to, 219
Imperialism. 228, 230, 248. *See also* Neo-imperialism
age of, 191
American, 191, 200–202
communist, 200, 202
conception of, 65
cultural, 69, 74
definition of, 73
development of, 67
economic characteristics of, 85
era of, 93
European, 78, 93
financial result of, 80
form of, 70, 191, 197
German, 201
global form of, 91
manifestation of, 67
national, 88, 90
Nazi, 201
neo-imperialism, 79, 83
new, 79
revival of, 96
Russian/Soviet, 206, 207
social, 74
superimperialism, 86

theory of, 82, 90, 91
ultraimperialism, 86, 90
warlike, 78
world, 191, 201, 202, 204, 206
Imperialist 79, 81, 85–88, 93, 95, 98
movement, 85
policy, 79
Incentive 149, 150, 155
Income 69, 81, 93, 116, 240
national, 103, 199
of workers, 121
pro capita, 120
source of, 119
India 31, 78
Industrialization 33, 40, 42, 45, 47, 48, 57, 58, 61, 246
advanced, 48
process of, 39, 47
Industry 3, 5, 145, 163
American, 181, 188, 196, 199
armaments, 176–178, 180, 183, 188, 193, 196
auxiliary, 182
British, 61
civil, 162
civilian, 176, 181
defence, 193, 198, 208
European, 188
financing of, 174
heavy, 68
military, 162, 163, 173, 177, 188, 190, 193, 195, 197
newborn, 58
private, 172, 196, 201
war, 172, 173, 181, 183
Inequality
create 237
reduce, 238
Infantino, Lorenzo 28
Inflation 196, 209
long-lasting, 171
major, 121, 122
monetary, 127
problem of, 126
wartime, 115, 182
Institute for Economics and Peace 255
Insurrection 7, 10
Interest 3, 9, 20, 22, 25, 29, 35, 39, 61, 66, 92, 119, 135, 153, 186, 206, 220
class, 67
common, 59, 65
economic, 58, 63, 65, 73
national, 73, 219, 225, 237
personal, 239
pursuit of, 219, 220
selfish, 71
special, 65
sphere of, 220, 239
International Association of Economic Sciences (AISE) 165
Intriligator, Michael 208, 255
Inuit 51
Investment 95, 108
boom of, 109
capital, 84, 95–97, 186, 187, 196, 207
foreign, 80, 95, 97
French, 96
increase in, 176
military, 176, 179, 180, 189
private, 95, 97
productive, 108, 116, 176
profitable, 81
state, 84, 95

Iraq 8, 105–107
Ireland 78
Iron Curtain 233
Isard, Walter 254, 255
Islam 229, 242
Israel 253
Istituto Nazionale di Finanza Fascista 121
Italian Union of Scientists for Disarmament 255
Italy 95, 99, 101, 120, 122, 125, 180, 182, 183, 218, 231, 235
Ivanov, Ivan 207

J

Japan 67, 69, 111, 115, 172, 175, 178–180
Jean, Carlo 217, 218, 230, 231, 235, 236
Jessop, Bob 161
Jevons, William Stanley 136
Jones, John Harry 102
Joxe, Alain 184

K

Kaldor, Mary 224, 227, 228
Kalecki, Michal 102, 176, 195
Kaplan, Robert David 227
Kautsky, Karl 82, 85–87, 90
Kaysen, Carl 6
Kennan, George Frost 191, 192, 211
Kennedy, Gavin 121, 135
Kennedy, John Fitzgerald 164, 195
Kennedy, Paul 113, 183, 196
Kerr, Prue 253
Keynesianism 185, 196
Keynes, John Maynard 5, 91, 99, 102, 106, 116–118, 179, 189
Khruschev, Nikita Sergeyevich 202–205, 209, 211
Kidron, Michael 187, 191, 210
Kissinger, Henry 9, 164, 241
Kjellén, Rudolf 229
Knies, Karl 62–64
Kosambi, Damodar Dharmananda 256
Krugman, Paul Robin 236
Krupp, company 199
Kuznets, Simon Smith 5, 103, 104, 111

L

Labour 86
- confiscation of, 114
- division of, 102
- forced, 114, 115
- productive, 110

Labour Party 120, 121
Lacoste, Yves 231, 232, 236, 242
Laissez-faire 17, 18, 21, 22, 28, 34, 40–43, 45, 46, 78
Landes, David Saul 184
Langer, William 97
Laplace, Pierre Simon 136
Latin America 198
Law 2, 14, 19, 21, 231
- code of, 31
- corn, 27
- economic, 36, 58
- international, 19, 21, 31, 34, 35, 48
- maritime, 65
- universal, 58

Lawrence, Klein 255

League of Nations 97, 125
Le Bras-Chopard, Armelle 50, 156, 162, 165, 248
Lemennicier, Bertrand 4, 7
Leninist 78, 88, 91, 93
Lenin, Vladimir Ilyich Ulyanov 67, 68, 82, 85–90, 203, 205, 248
Léon, Pierre 173, 174
Levi, Primo 256
Liang, Qiao 9
Liberalism 6, 68
Lincoln, Abraham 149
Lingotto, factory 182
List, Friedrich 2, 58–62
Lizza, Gianfranco 241, 247
Lloyd, David George 117
Lloyds of London 6
Loan 36, 96, 116, 117, 121, 178
  control over, 121
  enforced, 120, 121
  foreign, 122
  interest on, 116
  international, 118
Lockheed 241
Lorot, Pascal 230
Los Angeles 198
Loss 7
  economic, 103, 104
  huge, 124
  human, 103
Louis XV, King of France 19
Lundborg, Louis 108
Luttwak, Edward 155, 156, 206, 207, 218, 234–236, 242, 252
Luxemburg, Rosa 82–85, 89–91, 93, 98, 248

M
Macfarlane, Priscilla 103
Machiavelli, Niccolò 13–15, 17, 219
Macroeconomic 135
Madrid 165
Magdoff, Harry 191, 195, 196
Malacca Strait 6
Malthus, Thomas Robert 37, 38
Mandel, Ernest 185
Manhattan Project 144
Mann, Michael 50, 70, 74, 139, 161, 225, 226, 229
Market 136
  acquisition of, 80
  contraction, 4
  domestic, 67, 68, 79, 81, 83, 123, 124
  expansion of, 67
  external, 83
  financial, 82
  foreign, 82, 83, 123, 178
  free, 18, 27, 42, 92, 123, 124, 250
  greater, 234
  home, 248
  international, 58, 82
  national, 178
  new, 96
  non-capitalist, 83
  peaceful, 26
  wartime, 157
  world, 57, 88, 123, 191, 207
Marne 182
Marshal de Saxe: Sachsen Hermann Moritz von 248
Marshall Foundation 102
Marshall, George 136, 146, 158
Marshall, Monty G. 4

Marxism 2, 77, 140, 158, 159, 161, 245, 248, 250
Marxism–Leninism 200
Marx, Karl 77, 78, 89, 248
Massachusetts Institute of Technology (MIT) 198, 211
Massaua 99
Matière, Jean-Pierre 175
McCarthy, Terence 108
McCulloch, John Ramsay 18, 19, 25
McKean, Roland 134
Melman, Seymour 107, 112, 115, 126, 162, 163, 165, 194
Melon, Jean-François 19
Mercantilism/Mercantilist 1, 13, 17, 21, 22, 151, 154, 158, 174, 234, 235, 245. *See also* Neomercantilism
  age of, 65
  return to, 66
Mercier de La Rivière, Pierre-Paul 20–22
Michalet, Charles-Albert 91
Middle East 6, 126
Migration
  mass 247
Militarism 42, 43, 68–70, 72, 77, 79, 81, 84, 85, 88, 250
  activity of, 84
  advantage of, 43
  development of, 66
  European, 20, 30
  expanded, 79
  expansion of, 226
  interpretation of, 226
  state, 225, 226, 228
  thesis of, 226
  threat of, 226
Military-industrial complex 192–194, 199, 208, 210
Military strategy 252
Mill, James 30–32, 43, 81, 94
Mill, John Stuart 32, 33
Mills, Geoffrey 126
Milward, Alan Steele 42, 158, 172, 201
Mises, Ludwig von 68, 69, 92, 93, 99, 109, 110, 126
Molinari, Gustave de 71
Monopoly 78, 85, 88–90, 191
  capitalist, 90, 172, 185, 186, 201
  dominance of, 89, 90
  state, 172, 201
Monopsony 178
Monroe Doctrine 191
Montchrestien, Antoine de 16, 17
Montesquieu, Charles-Louis de Secondat de la Brède et de 18
Moreau Defarges, Philippe 229–234, 242
Moresco, Emanuel 102
Morgenstern, Oskar 136–138, 145, 164
Morgenthau, Hans J. 3, 92, 219, 220, 223, 224
Moszkowsky, Natalie 186
Moulton, Harold Glenn 102
Movimento Federalista Europeo 125
Mozambique 207
Mueller, Dennis Cary 239
Mueller, John 6
Munich Soviet Republic 69
Mun, Thomas 16
Mutually Assured Destruction (MAD) 150, 254

N

Nagasaki 165
Nahum, Mordo 256
Nanni, Walter 11
Napoleon First, Emperor of the French 65
NASA 195, 198
Nash, John 138
Nation 139, 149, 165
 hostile, 140
 neighbouring, 140
Nationalism 72, 94, 98, 246
 British, 80
 German, 201
NATO 202
Nazism 172, 201
Negotiation 153, 154, 157
Neo-imperialism 68
Neoliberalism 4
Neomercantilism 58, 66, 67
 nineteenth century, 59
Neorealism 3, 222–224, 250
 structural, 218–221
Netherlands 43, 62, 202
Network of European Peace Scientists 255
Neumann, John von 137, 138, 144
Neurath, Otto 69
New York 255
Nietzsche, Friedrich 4
Nixon, Richard 206, 241
Nobel Prize 138, 148, 149, 152
Nordhaus, William 111
North Atlantic Treaty 218
North, Douglass 251
North, Dudley 19

O

Olivetti (company) 199
Optimum
 individual 138
 social, 138
Oslo 254
Ottoman Empire 98
Overabundance 246, 248, 249
Overproduction 81–83, 86, 88, 98, 186, 187

P

Pacific Architects & Engineers (PAE) 241
Pacification
 elements for 57
 international, 62, 65
 process of, 60, 64
 tendency towards, 64
Pacifism 57, 64, 74, 177
 modern, 74
Palestine 253
Palmieri, Giuseppe 23
Pareto, Vilfredo 44, 146
Passy, Frédéric 41
Patalano, Rosario 210
Pavanelli, Giovanni 121, 122
*Pax Americana* 217
Peace 1, 2, 4–6, 150, 157, 158. *See also* Pacification; Pacifism
 chances for, 224
 future, 46
 generalized, 29
 idea of, 66
 international, 20, 22, 24–26, 33, 42, 124, 254
 lasting, 63, 64, 123, 204, 222
 long, 221, 223

long-term, 254
maintenance of, 150, 208
periods of, 221
permanent, perpetual, 26, 31, 42, 63, 115, 188, 189, 249
research, 156
science of, 156, 254
social, 64, 173, 188
undesirable, 189
universal, 61
work of, 68
world, 219, 229
Peace Science Society International 254
Peacetime 103, 104
Pentagon 143, 163, 187, 195, 197–199, 226, 253, 254
capitalism, 163
Petersen, Howard Charles 206
Peugeot, company 182
Philippe d'Orléans 19
Physiocracy 20
Pierpont Morgan, John 79
Pigou, Arthur Cecil 114, 116
Pizzorno, Alessandro 139
Poland 176
Polemology 249
Politics
American 235
form of, 219
globalized, 237
idealism in, 219
international, 219, 220, 222, 223, 235
understanding of, 220
world, 221, 235
Portugal 97
Possony, Stefan 174, 210
Power 2, 5, 6, 14–16, 21, 24–27, 29, 31, 33, 34, 43, 45, 46, 48–50, 136, 142, 152, 155
accumulation of, 94
balance of, 208
capitalist, 83, 85, 89
coercive, 125
destructive, 174, 189
dominant, 221, 223
economic, 188, 194, 196, 199, 206
excessive, 223
external, 218
instruments of, 67
internal, 218
labour, 186
leading, 150
maritime, 15, 16
military, 183, 188, 193, 194, 206, 226
misplaced, 193
national, 180, 183
nuclear, 152
of state, 112
political, 13, 25, 40, 49, 66, 72, 82, 93, 97, 217, 221, 233
principle of, 65
productive, 61
purchasing, 83–85, 102, 118, 119
relations of, 90
rivalry of, 231
science of, 229
seizure of, 203
struggle for, 219
Price
construct 103
evaluate, 103
reduction in, 123
revision of, 126

rise in, 119, 120
share, 107
Princeton 148
Prisoner's dilemma 150
Product
final 103
finished, 118
intermediate, 104
national, 103
war industry, 103
Production. 107. *See also* Overproduction
anti-economic, 102
armaments, 68
arms, 241
capitalist, 64, 77, 79, 83
capitalist-industrial, 93
civil, 107, 108, 162
concentration of, 85, 89
control over, 122
convert, 174, 182
domestic, 59, 83, 96, 225
downsize, 174
economic, 112
excess of, 93, 123, 171, 248
fall in, 63
future, 108, 112
increase in, 116, 158, 171, 176, 182, 203
industrial, 145
interruption of, 228
means of, 85, 186, 188
military, 162, 174, 176, 180, 182, 184, 190, 199, 207
mode of, 74, 77, 83, 90, 225, 228, 245, 248
of commodity, 85
of peace, 77
of war, 77
outlet for, 248
peacetime, 104, 184
process of, 110
rationalization of, 70
reduction of, 57, 108
rise in, 81
state, 225, 234
surplus of, 83, 86, 123
war, 250
wartime, 103, 174, 184, 190
Profit 21, 32, 42, 44, 151, 154, 163, 174, 182, 187, 188, 199
control of, 178
distribution of, 151
repatriation of, 186
wartime, 173, 174, 182
Proletarian revolution 84, 88, 91
Proletarization 77
Propaganda 194
Prosperity 171, 172
apparent, 109
economic, 26, 47
general, 22, 45
idea of, 111
material, 41
national, 31, 41
peaceful, 47
period of, 109
real, 110
wartime, 109, 111, 115
Protection 62, 65. *See also* Protectionism
customs, 62
period of, 58
Protectionism 25, 37, 44, 58, 60–62, 67, 82, 92, 96
give rise to, 60
introduction of, 63
return to, 66

Public Choice Theory/School 10, 238–241
Putilov (company) 199

Q

Quade, Edward S. 134, 135
Quesnay, François 22, 23, 34, 59

R

Radnitzky, Gerard 145
Rand Corporation 135
Rapoport, Anatol 140–144, 253
Ratzel, Friedrich 230, 231
Reagan, Ronald 247
Realism 3. *See also* Neorealism
  principle of, 219
  structural, 218–221
  theory of, 218
Realist school 92
Realpolitik 219, 223
  traditional, 218
Rearmament 1, 157, 163
  competitive, 176
  massive, 176, 183
  maximum effect of, 176
  policy of, 172, 186, 191
  rapid, 171, 172, 191
Regime
  monopsonic 161
  semi-monopolistic, 161
Renault, company 181, 182
Reproduction 186
Resource(s)
  current 116
  economic, 189, 190, 209
  exploitation of, 197
  financial, 120, 173, 190
  future, 116
  human, 190
  mineral, 247, 250
  monetary, 116
  natural, 172
  productive, 119
  strategic, 247
  use of, 103, 121, 173, 175, 206
  useful, 250
  waste of, 189
  water, 247
Restrictionism 92
Rhodes, Cecil 99
Ricardo, David 25, 35–37, 114
Ricard, Serge 9, 182
Richardson, Lewis Fry 139, 140, 164
Robbins, Lionel Charles 91, 92, 98, 102, 159, 160
Robinson Crusoe 136
Robinson, Joan 165, 175, 176, 197, 253, 254
Robinson, Ronald 28, 51
Rockefeller, John Davison 79
Rockoff, Hugh 126
Rodbertus, Johann Karl 98
Romania 97
Roosevelt, Theodore 9, 79, 241
Roscher, Wilhelm 62, 63
Rosemberg, Samuel 210
Rosovsky, Henry 179
Rossi, Pellegrino 41
Russett, Bruce Martin 107
Russia 64, 66, 67, 95, 97, 115, 118, 180, 206, 209

S

Sachsen, Hermann Moritz, von 50
Said, Edward W. 242

Saint-Simon, Claude-Henry de Rouvroy, comte de 43, 45, 46
Salter, Arthur 97
Samuelson, Paul Anthony 162
San Diego 198
Sassoon, Donald 165
Savary, Jacques 18
Savona, Paolo 218, 231, 235
Saxe, Marshal. *See* Sachsen, Hermann Moritz, von
Say, Jean Baptiste 38, 114
Schelling, Thomas 142, 147, 148, 152–156, 164, 165
Schmidt, Christian 127, 134, 136, 137, 158, 164, 165, 249
Schmoller, Gustav 62, 64
Schneider-Creusot, company 199
Schumpeter, Joseph Alois 67, 69–73, 75, 146
Schwartz, Anna Jacobson 126
Schwartz, Harry 206
Scientists for Peace 255
Seattle 198
Sector 57, 60, 68, 72
  armaments, 104
  business, 108
  civil, 108, 110
  defence, 106, 108, 109
  financial, 217
  industrial, 57
  military, 60, 110
  peaceful, 228
  private, 113, 121
  productive, 233
  public, 105
  strategic, 234
  welfare, 107
Security 180, 191, 193, 194, 196, 221–224, 234
  territorial, 234
Sedillot, René 182
Segreto, Luciano 177, 183–185
Selten, Reinhard Justus Reginald 155
Sen, Amartya 139, 241, 255
Shapiro, Ian 241
Sharon, Ariel 164
Shaw, Martin 6, 88, 89, 188, 224–226
Shipbuilding 172, 179, 183, 199
Shortage 246, 248
  material, 8
  of land, 248
  oil, 8
  perception of, 249
  water, 8
Shotwell, James Thomson 101
Shubik, Martin 145
Silberner, Edmund 64, 161
Simoncelli, Maurizio 10
Singer, company 199
Singer, David J. 139
Sismondi, Charles Léonard Simonde de 83
Situation 136, 138, 142, 143, 145, 158
  bargaining, 153
  bipolar, 218, 223
  changing, 218
  conflict, 144, 153
  crucial, 223
  international, 221, 222
  multipolar, 223
  state-centric, 227
Skidelsky, Robert 121
Small, Melvin 139
Smith, Adam 2, 24, 35, 36, 40, 59, 114, 134, 250
Smith, Wayne 135

Socialism 77, 84, 86, 87, 90, 92, 200–203
Society 13, 23, 35, 36, 38, 42, 45–48, 51
anarchic, 237
bourgeois, 70
capitalist, 70–72
civil, 48
contemporary, 42, 46, 51, 71, 225, 226
free, 239
global, 224, 237
industrial, 45–48, 71
liberal, 68
open, 237
preindustrial, 69
present-day, 71
task of, 70
Sociologist 33, 45, 48
Solow, Robert Merton 236
Sombart, Werner 27, 64–68, 74
Sorokin, Pitirim Aleksandrovich 139
South Africa 78
South America 30
Soviet Block 200–202
Soviet Union 6, 113, 144, 151, 154, 185, 187, 194, 200–202, 204–207, 210, 211, 223, 247, 253
Space vital 123
Spain 23, 97
Spence, William 30
Spending
armaments 176, 189, 190, 194, 206
industrial, 176, 197
war, 174, 176, 187, 194, 199, 206
Spengler, Oswald 4
Spin-off 160, 162, 176, 178–181, 184
military, 179, 181, 182
wartime, 104
Staley, Eugene 95–97, 159
Stanford University 95
State
absolutist 65, 73
affair of, 16
behaviour of, 73, 219, 221, 230
British, 226
defeated, 14, 18
dominant, 223, 228
egoism of, 17
enrichment of, 14, 16
European, 66, 235
formation of, 224
German, 58, 62, 66
greatness of, 14
imperial, 228
importance of, 234
interest of, 14, 23, 24, 40, 46, 65, 66
intervention, 57, 58, 63, 69
militarization of, 66, 67
modern, 65
nation, 98
national, 59, 63, 225, 234
need for, 16, 22
newborn, 14
organization of, 221, 226
permanence of, 58
powerful, 218, 237
power of, 22, 220, 221, 223, 226, 229, 233
problem of, 14, 50
prosperity of, 235
security of, 21
sovereign, 227, 233

strong, 64, 67
survival of, 218, 227, 233
system of, 221, 224, 225, 236
wealth of, 14, 16–18, 22, 25, 32, 33
Stiglitz, Joseph 105–107
Strategy 3
conflict, 153
dominant, 138
interdependent, 138
international, 153
military, 135, 143, 155
war, 148, 155
Strazzari, Francesco 11
Stuart, Reginald 9
Supply 187, 189
Surplus 248, 249
monetary, 122
outlets for, 123
taxation of, 122
Sussex 117
Swedish Academy of Science 164
Sweezy, Paul 184, 190, 191
Switzerland 43, 62, 67
System 136, 139, 143, 148, 152, 155
anarchic/anarchy of, 221, 228
assistance, 105
capitalist, 194, 205, 209, 224
diplomatic, 157
distribution, 70
economic, 60, 62, 101, 173, 187, 203, 204
expansionist, 191
global, 224
imperialist, 196
industrial, 173, 194, 198, 207
international, 221, 222, 224, 228
interstate, 224
military-industrial, 207
national, 59, 221
of taxation, 125
of war, 114
permanent, 62
political, 106, 222, 224
production, 70, 108
productive, 182
protectionist, 60
soviet, 194
state, 221, 224, 225, 236
trade, 62
war, 157, 187, 189
world economic, 218

T

Tax 84
burden, 125
income, 114
rate, 112
special, 125
wartime, 112
Taxation 25, 30, 36, 37, 149
excessive, 140
future, 116
heavy, 102
increase of/rise in, 119–121
level of, 122
low, 121
percentage of, 150
progressive, 116
special, 122
war taxation, 38
wartime, 102
Technology
advanced 177, 179–181, 183, 197, 199, 208
civilian, 179–181, 183–185, 196

industrial, 172, 175, 177, 183, 185, 196
military, 177, 180, 181, 183, 184, 190, 197, 208
transfer of, 177, 179–181, 183, 184, 210
war, 172, 177, 179, 181, 184
Temple, William 14
Territory 218, 227–229, 231, 232, 234, 236
national, 217
possession of, 230
problem of, 231
Tertrais, Bruno 11
Theorem
mathematical 137
Minimax, 137
Theory 135, 137, 138, 140, 141, 144, 147, 153, 155, 157, 158, 250
decision, 239
economic, 238, 240
economic theory of conflict, 251
Game, 133, 135, 137–139, 141–145, 147, 148, 150, 152, 153, 156, 157, 164, 239, 252
neorealist, 222
Public Choice, 134, 150
Rational Choice, 147, 150
Thierry, Augustin 46
Thompson, Edward Palmer 194
Thucydides 219
Tobin, James 111
Tobin, Jim 149
Tooze, Adam 172
Trade 78, 80, 82, 86, 93, 97, 136, 137, 139
aggressive, 1, 16, 19, 21, 28
British, 80
community of, 66
competition, 154
domestic, 21, 37
expansion of, 19, 33
foreign, 1, 16, 17, 19, 21, 25, 27, 30, 64
free, 17–22, 24–26, 28–30, 32, 33, 35, 40–44, 49, 57, 59, 61, 62, 66, 82, 86
growth in/increase in, 23, 30
international, 16, 18, 20, 21, 24, 26, 30, 32, 57, 60, 61
jealousy of, 24, 29, 50
monopolistic, 25, 32
mutually advantageous, 27
pacific, 1
privileged, 30, 32
restriction, 140
road of, 66
sweet/ sweetness of, 18, 25, 30
wars, 16, 19, 24, 27, 29, 30, 46
world, 139
Trade balance 16, 17
Trade Unions 120
Treaty 59
Zollverein, 59
Trebilcock, Clive 104, 160, 162, 176–181, 183, 184
Trebilcock, Michael J. 228
Treitschke, Heinrich von 48
Tucker, Josiah 27, 28
Tullock, Gordon 134, 150–152, 238–241, 245, 246
Turkey 202
Twin Towers 8

U

Underconsumption 83, 90, 98

Underdevelopment 57
Underwood (company) 199
Unemployment 171, 174, 176, 187, 204, 209
  mass, 176
UNESCO 175, 256
United Kingdom (UK) 202
United Nations 158
United States Army Air Force 135
United States Atomic Energy Commission 144
United States (US) 4, 6, 8, 9, 58, 62, 64, 81, 98, 105, 106, 108, 111, 113, 115, 117, 123, 144, 151, 153, 154, 163, 171, 176, 183, 185, 187, 188, 191–197, 199, 201, 202, 205–207, 209–211, 223, 228, 229, 236–238, 241, 253
University of California 145, 198
University of London 159
University of Michigan 156, 157, 189
University of Paris 158
University of Princeton 164
University of Yale 145
Usher, Dan 147
US Steel 198
Utility 228
  collective, 134
  individual, 134, 136
  maximization of, 136, 239
  principle of, 239
  social, 134

V

Vaïsse, Justin 9, 224, 229
Vanderlint, Jacob 19
Veblen, Thorstein 159
Verdun 182
Verri, Pietro 18
Versailles 201
Versailles Conference 117
Vickers-Armstrong, company 199
Victory 150, 151, 160, 161
Vienna Circle 69
Vietnam 162, 164, 181, 191, 193–195, 206
Violence 134, 145, 147, 150–153, 251
  exercise of, 235
  forms of, 73
  fragmentation of, 227
  internal, 105
  monopoly of, 126, 218, 225, 227
  of arms, 65
  organized, 227
  principle of, 68
  private, 240
  proliferation of, 217
  public, 240
  recourse to, 65
  sharpening of, 217
  spread of, 71
  use of, 69
  warlike, 71
Von Neumann, John 253

W

Wallerstein, Immanuel Maurice 242
Wallis, John Joseph 251
Walras, Léon 44, 136, 149, 252
Waltz, Kenneth Neal 98, 218, 220–224, 241
War. *See also* Warfare; Wartime, warlike

act of 73
aerospatial, 6
African, 7
alarm of, 80
Anglo-Boer, 78, 79, 91, 115
armaments for, 172, 178, 179, 184, 186, 187, 189, 190, 210
art of, 23, 36, 50, 77
asymmetrical, 7
atomic, 252
benefits of, 171, 173, 181, 182
burden of, 36
cause of, 60, 140, 146, 159, 200, 221, 226–228
Civil War, 7, 10, 14, 15, 17, 34, 159
Cold War, 4, 49, 109, 111–113, 135, 138, 143, 150, 171, 184, 185, 190, 192–194, 197, 199, 200, 203, 211, 227, 234, 235, 253
colonial, 248
conduct of, 114
consequences of, 1
contemporary, 228
continuity of, 237
cost of, 4, 24, 34, 37, 38, 42, 51, 60, 107, 116
damage of, 107
economic, 171–177, 187, 189, 208
economic policy of, 147, 159, 160
economic role of, 172, 174, 176
effects of, 18, 31, 32, 36, 103, 125
electronic, 6
elimination of, 28, 42, 44, 48
emergency of, 118
Eritrean, 6
Ethiopian, 6
ethnic, 7
expenditure on, 226
expense of, 61
expensive, 114
explanation for, 218, 222, 231
Falkland/ Malvinas War, 6
First Gulf War, 6
First Opium war, 78
First World War, 5, 48, 68, 69, 78, 86, 97, 102, 109, 117, 122, 125, 171, 173, 177, 179–183, 190, 199, 200
Franco–Prussian, 43
frequency of, 222
function of, 73
global, 7
Gulf War, 106, 150, 241
horror of, 253
ideological, 7
impact of/incidence of, 221, 225
imperialist, 186, 204, 208, 228
implication of, 253
infrastate/interstate/suprastate/non-state, 6, 7
intensity of, 139
intercontinental, 62
irrelevance of, 133
Korean War, 112, 150, 154, 175
legitimacy of, 64
lies of, 256
local, 7, 235
logic of, 24, 117
mobilization for, 119
modern, 60
motivations for, 147
Napoleonic, 30, 36, 38, 65
necessary, 14, 37, 43

new, 228, 231
non-military, 9
nuclear, 143, 151, 165
obsolescence of, 74
of aggression, 64
of self-defence, 68
Opium War, 249
organization of, 225
origins of, 222
outbreak of, 63, 103, 149, 152, 228, 235
pay for, 110
permanence of/persistence of, 43, 218, 224
permanent, 187, 193
policy of, 147, 159, 160
possible, 62
pre-modern/postmodern, 7
preparation for, 39
preventive, 15, 37, 64
proliferation of, 158
psychological, 7
rational, 133, 150, 152
rationality of, 226, 251, 253
reasons for, 24, 25, 27
reconversion of, 69
religious, 7
reparations, 110
Second Anglo-Boer War, 2
Second World War, 3–5, 33, 49, 91, 101–104, 106, 109, 110, 112, 118, 122, 137–139, 171, 172, 175, 185, 186, 196, 199, 201, 210, 223, 230, 238, 240, 253, 256
state, 65, 69, 74
strategy, 138, 148, 155
subversive, 7
success of, 24
threat of, 153, 235
total, 103
trade, 16–19, 22, 24, 27, 29–31, 35, 37, 47, 137
transnational, 7
Vietnam War, 107–109, 181, 191, 193–195, 206, 241
world, 70, 71, 74, 101, 118, 123, 124
Warfare 185, 225, 227, 234, 235
Warlord 250
Warsaw Pact 210, 218
Wartime, warlike 14, 15, 26–28, 31, 33, 35–37, 40, 44, 173, 174, 181, 182, 184, 190
Waste 249
Watergate Scandal 206
Water, Peace and Security (WPS) 256
Wealth 13, 14, 17, 20–23, 25, 29, 30, 32, 34–36, 39, 40, 46, 50, 80, 88, 93, 94, 136, 141, 146, 152, 157, 183, 217, 228
distribution of, 59, 94
domestic transfer of, 152
increase in, 61, 69, 70, 80, 93, 110
loss of, 63
material, 61
mobilization of, 122
national, 23, 25, 46
production of, 146
pursuit of, 70
redistribution of, 36
social, 146
surtax on, 125
unlimited, 197
Weber, Alfred 66
Weber, Max 49

Weidenbaum, Murray Lew 196
Weingast, Barry Robert 251
Welfare 189, 192
  increase in, 104, 109
  substitute for, 111
West 236, 237
West Germany 175
Wicksell, Knut 239
Wilhelm II, king of Prussia 124
Williams, Michael 224
Wilson, Peter A. 127
Wilson, Thomas Woodrow 9, 241
World Commercial Court 97
World Investment Bank 97
World Investment Commission 97
Wright, Quincy 44, 97, 140, 141, 158

X
Xiangsui, Wang 9

Z
Zero-sum game 2
Zhdanov, Andrei Alexandrovich 201, 202
Zimmerman, William 200

The manufacturer's authorised representative in the EU is Springer Nature Customer Service Centre GmbH, Europaplatz 3, 69115 Heidelberg, Germany. If you have any concerns regarding our products, please contact ProductSafety@springernature.com

Printed and bound by CPI Group (UK) Ltd, Croydon, CR0 4YY
30/09/2026
02225644-0007